W9-BRU-735

Alaska
The Last Frontier

Books by Bryan Cooper

Alaska
The Last Frontier

Bryan Cooper

William Morrow & Company, Inc.
New York 1973

Published in the United States in 1973.

Published in Great Britain in 1972.

Printed in the United States of America.

Cooper, Bryan.
Alaska: the last frontier.

Bibliography: p.
1. Petroleum industry and trade—Alaska. 2. Alaska—Economic conditions. 3. Alaska—Description and travel. I. Title.
HD9567.A4C66 330.9'798 78-183955
ISBN 0-688-00016-9

Acknowledgments

My thanks are due to many people who helped in the preparation of this book, particularly Julius Edwardes, John Collins and Alwyne Thomas of British Petroleum, and the Company itself for providing facilities to visit the North Slope; Dr Victor Fischer of the University of Alaska; Joe Fitzgerald of Atlantic Richfield; C. V. Chatterton of the Standard Oil Company of California; John Borbridge of the Alaska Federation of Natives; and Dr Max Brewer of the Naval Arctic Research Laboratory.

Alaskan hospitality is boundless and I am personally indebted to the kindness shown to me by both old and new friends alike in Alaska, especially Charles and Pamela Towill, Art Davidson, Joe and Claire Fejes, Betzi Woodman, Joe and Donna Rychetnik, Don Sheldon, and Richard and Jean Montague.

I also wish to thank Maurice Cochrane of BP and Humble Oil and Atlantic Richfield for some of the photographs used in this book and the Petroleum Press Service for their kind permission to reprint the diagram of North Slope leases.

Finally, my deepest gratitude to my wife Pat for her unfailing support and encouragement over the years and to our children, Jeremy, Jane, Annette and David, for their cheerful tolerance when a sometimes bearish father demands quiet.

B.C.

Contents

PART THREE

THE ENVIRONMENT

The photographs (between pages **112** *and* **113***) are by the author except where otherwise stated.*

MAPS AND CHARTS

PART ONE

The Country

I

Prologue

Alaska is many things to many people. To those who seek only its wealth it is a forbidding land of ice and blizzard, of long sunless winters and unendurable cold. To the hunter it is a land still rich in wildlife—caribou, moose and bear—in which a man can pit himself against nature and kill his own meat. Unless it is merely a trophy he wants or the forty-dollar bounty which is still paid for killing wolves. To others it is a wilderness of purity and beauty where rivers run crystal clear and hills rest in solitude like the humped backs of buffalo and even the bleak Arctic plain blossoms with a carpet of poppies and wild flowers during the two months of summer. The winding Yukon, with its ghost towns and pine forests and lonely miners still panning for gold, continues to cast its spell. Alaska is a state of mind, existing in the dreams of many who might never see it for themselves but who are enriched because it is there. As Alaskans say, when you come to Alaska you are 'inside'—anywhere else is 'outside'. It is the difference between reality and dream.

This book is partly about the great Alaskan oil rush, following the discovery on the North Slope of one of the world's largest oilfields and by far the biggest ever found in the United States. But it is also about Alaska itself and the impact of that discovery on a land that has lived for so long in isolation. There have been many oil booms before, at other times and in other places, but none have aroused such passionate controversies as that which began in Alaska in 1968, only to be abruptly halted for a time as a result of campaigns fought by conservationists to protect the wilderness from the encroachment of industry. For

Alaska is one of the few areas of wilderness left in the world. It is the last American frontier.

The plane from Seattle hugs the narrow coastal strip of south-eastern Alaska on its way to Anchorage. Great snow-streaked mountains rise steeply from the sea, ribboned with massive glaciers oozing into the North Pacific. Hidden in shadow from the westering sun we have been chasing all day are deep gorges and inland waterways, twisting between the islands. It is difficult to tell where the islands end and the mainland begins. Over to the east, only thirty miles away, is Canada. This part of Alaska seems barely to have a foothold on the North American continent. It is so narrow and mountainous that there is nowhere to build a road connection with the rest of the state or between the towns themselves. Even Juneau, the capital of Alaska, is cut off except by air and sea. It seems strange that the border with Canada isn't further inland, and then you remember that this land once belonged to Russia, who sold it to the United States just over a hundred years ago because she feared the British might take it over. But for the vagaries of history, and a time when countries bought and sold huge chunks of territory like real-estate dealers, Alaska would belong to Canada. It would make sense geographically. As it is, the boundary with Canada is a tribute to some canny bargaining between Scotsmen of the Hudson's Bay Company and the Russians.

Turning westwards across the Gulf of Alaska, the plane skirts the even more massive Chugach Mountains. These snow-capped peaks were the first view that Vitus Bering had of this vast land when he made the first recorded discovery of Alaska over two hundred years ago. North from here lie 800 miles of mountain ranges and valleys and rivers to the icy plains of the North Slope bordering the Arctic Ocean. There are some areas yet to be explored. It was an exciting moment for Bering's crew, and it still is for the modern traveller. This is the last frontier, where the sense of adventure is still very real. The land of the Eskimo and the Yukon gold-miner, the caribou and the grizzly bear. A land one-fifth the size of the rest of the United States, bigger than Britain, France, Germany and Italy combined, with a population of only 280,000, less than the Tyneside city of Newcastle or Tulsa, Oklahoma.

The other passengers feel it too. The two hunters from Texas, sporting check plaid jackets and fur caps and shiny new gun cases. 'Reckon it's worth three thousand bucks to get a polar bear,' one says. 'They say there won't be any more left soon.' The tourists get together in the forward lounge, comparing itineraries. 'You going to Fort

Yukon?' Two middle-aged couples from the East have latched on to each other for companionship in an unknown land. They pair off naturally, the men on one side and the women on the other. 'Maybe we could all go up there together. There's a stockade and Indian village, just like the old days.' The salesman returns from an unsuccessful flirtation with the stewardess. 'Yeah—this state is wide open.' He picks up where he left off. 'They've got the money to spend—all this oil, you see. And we've got the goods for 'em to spend it on.' He's a man with a ready laugh and tired eyes and a conversation sprinkled with stories about the girls he has known on his travels. Now he has come to take charge of a new sales area in Alaska. A plum ready for the picking. His company sells anything that anybody wants, from insurance to canned tomatoes.

At the back of the plane a group of oilmen exchange news of old friends and rates of pay. They are big, paunchy men from Texas and Oklahoma. 'Hank? He was in Indonesia last time I heard.' 'Yeah—the play's pretty big out there.' One of them has just come back from the Middle East and hasn't been to Alaska before. He's a driller. 'They've offered me five hundred a week to go up to the North Slope.' 'You'll earn it.' 'Rough?' 'And some.' Unexpectedly, they are not drinking hard. They nurse the glasses of bourbon in their big hands until the ice melts. 'No point in getting too much of a taste for it,' they tell the newcomer. 'All the camps are dry.' The driller is used to prohibition from his days in Saudi Arabia, but is surprised to hear it applies to the North Slope as well. 'Not even beer?' 'No.' 'How come?' 'They reckon you might get liquored-up and wander off the camp. That's one hangover you wouldn't wake up to. You wouldn't wake up. Here—read this.' It is a booklet on how to stay alive in the Arctic. What to do in case of severe chilling, frostbite, trench foot and snow-blindness. They hand it out to the oilmen in the camps. The driller looks sceptical. 'Couple of guys in my outfit got caught in a blizzard last month,' drawls the cat-skinner from Dallas. 'One of them lost three toes.' The driller is convinced. There seems more to the North Slope than drilling for oil. Like just surviving.

The pilot drops altitude over the sea as we near Anchorage. There are more mountains ahead. They arc majestically round the head of Cook Inlet and the river valley beyond. Then the plane touches down at Anchorage airport. This, one feels, is where civilisation ends and the land of the pioneer begins. Behind are air-polluted cities and scarred landscapes and suburban sprawls. The strikes and tensions of London and New York. Ahead is the wilderness, one of the few areas in the world still virtually untouched by man. Or is it?

A 747 jumbo-jet from Tokyo has just landed and is disgorging its 400 passengers. One is reminded that the rapid development in international air travel has put Anchorage at the crossroads of the world, a midway stepping stone in the polar route between Europe and the Far East. But most of these passengers will just make a short stop-over —there are expected to be as many as two million through passengers a year by 1985. Only a few will stay for any length of time. And in the main airport building it seems that a large proportion of these are camera crews. Black leather boxes of film equipment are scattered everywhere. Television units from Germany and New York are just arriving, others are on their way out. A bearded BBC producer is shepherding his crew towards the Customs area while an Italian argues excitedly about some missing cases of film. Alaska has been discovered by the restless, prying eye of the documentary film camera. One feels sympathy for the men who have to haul around so much heavy equipment. They seem to spend most of their time in air-freight offices and filling in Customs forms.

'So you're going to write a book on Alaska.' Greeting me at the airport is Charles Towill, an old friend from New York days and now public-relations manager in Alaska for British Petroleum. I had wondred how Charles, an Englishman from Devon who fell in love with New York and lived there for the past fifteen years, more of a New Yorker than many native-born, had taken to his new assignment in Alaska. 'It's a great place. Fantastic. So much room to move and breathe in. No smog—no traffic jams. We're buying a house, up in the hills there. The caribou come right down to the garden.'

His enthusiasm echoes the yearning of other friends in New York who had spoken of 'getting the hell away from all this' and emigrating to Alaska. That it how Americans think of Alaska, in much the way that digruntled Britishers consider emigrating to Australia. Alaska might belong to the United States—it was the forty-ninth state to join, in 1959—but in many ways it is an anachronism. It is separated geographically from the rest of the country by 600 miles and mentally by a hundred years or so. That was when the last of the old frontiers vanished in California and Texas. In Alaska the frontier is here and now. The rest of the continental United States is the 'lower forty-eight'. It might be another country to hear Alaskans talk.

And this enthusiasm I was to find reflected by nearly everyone I met in Alaska, whatever their views on how the state should be developed or even whether it should be developed at all. One is never in any doubt what those views are. There are two basic essentials about Alaskans. They speak their mind, and they have the natural friendliness and hospitality of people living in small communities where everyone seems to know everyone else. Visiting Alaska is like one of those snowballing chain letters. Each person you meet puts you on to half a dozen others, and so on. I hadn't been in Anchorage for more than a few hours before I had introductions to people all over the state, from politicians in Juneau to retired gold-miners in Fairbanks, from Eskimo leaders in Barrow to seal-hunters in Nome.

I had intended to write solely about oil in Alaska, the great discovery of 1968 at Prudhoe Bay which has led to one of the biggest oil rushes in the world since the boom years in the Middle East. It was a dramatic enough event, dwarfing the great gold rush of seventy years before which was the last time Alaska hit the world headlines. The scramble for leases, with companies playing cloak-and-dagger games of industrial espionage—the biggest lease sale in history, resulting in a bonanza of nearly a billion dollars for the state at a time when its total income was a fifth of that amount—the sheer immensity of the task of bringing oil equipment to the Arctic wilderness and of men battling against blizzards and temperatures of sixty below to keep the drill-rigs working. Boardroom decisions taken in London and New York to pour millions into the search with no guarantee of getting any of it back, some companies failing and others succeeding beyond their wildest hopes. And the reasons which brought this most powerful of modern industries to Alaska in the first place.

But this is not the whole story, as I was to realise after two months and several thousand miles of travelling throughout the state. In going to Alaska in the first major search for oil in Arctic conditions the oil companies have found themselves up against something unique. It is not just the technical problems involved, great though these are. The companies are used to difficult conditions, whether they be in desert, jungle or under the sea, since it is in the nature of oil that it is often found in the most inaccessible places. They have the scientific and financial resources, allied to imaginative flair, to overcome these. But in Alaska the problems are human and emotional. They have to do with the nature of its people. To the great surprise of the oil companies, having succeeded physically against a formidable environment, they suddenly found themselves in the middle of such controversies as conservation and wildlife preservation, the claims of Eskimos and Indians after years of neglect, political hassles between state and federal government, and the whole question of the quality of life and the price that should or should not be paid for industrial progress. All these conflicts existed before, but oil is the catalyst that has brought them prominently into the open.

On one of my last evenings in Alaska I sat at the bar of the Baranov Hotel in Juneau listening to a group of politicians argue about what to do with the $900 million the state had harvested from the sale of leases. It is what most people talk about in Juneau, an endless subject of debate in the state legislature. Having the money and not being able to agree on what to do with it has caused more bitterness than ever existed in the days when there wasn't any and the state had to rely on federal grants to rescue it from annual bankruptcy. It was raining as usual outside, the kind of soft rain you get in Ireland. It rains so much in Juneau that the buildings are built out over the sidewalks so that people can move about without getting soaked. Saloons like the Red Dog and the Top Hat Bar were warming to the night's business. Happy hour, when drinks are half-price to get customers in the mood to carry on drinking afterwards. There isn't much else to do in Juneau anyway.

'It's no goddam use sitting on all that money,' one of the politicians was asserting loudly.

'We're getting the interest from it,' another argued. 'There's a lot we can do with that.'

The money has been invested in government securities and the interest comes to around $194,000 a day. The big argument is whether to spend just that or eat into the capital.

'Peanuts. What this state needs is development. Build roads to attract more industries in—then we can really get moving.'

'Like California,' another said. 'That's why I left to come up here.'

'Keep Alaska the way it is—that what you want?'

'The pollution down there . . .'

'All this talk about pollution. I'm sick to death of outsiders screwing up their own cities and then coming here to tell us how to run things.'

That is the crux of the matter. Should Alaska be kept as a wilderness, one of the few that remain in the world? After all, that was why most people came up here in the first place, to escape industrialisation. Or should it be developed to raise the living standards of its people. And if so, what kind of development? And how could one be sure it was Alaskans who benefited and not just those from outside? Some Alaskans had got rich from the oil companies, mostly those running their own businesses or otherwise alert to the main chance, but the big salaries were going to specialists brought in from the other states. From what I had seen, the ordinary Alaskan hadn't got much out of it —the reverse, in fact, because the influx of people with money to spend had merely served to push prices up. A short taxi ride in Anchorage can cost six dollars, bread is about sixty cents a loaf, a bottle of beer one dollar.

'Anyway, there isn't much can be done until this business with the natives gets settled.'

Yet another problem. The man who spoke was referring to the bills then being put to Congress to settle the question of Eskimo and Indian land rights and the amount of compensation that should be paid to the natives for land taken away from them. It should have been done a hundred years ago, but no Congress ever got around to it. Until the matter was settled, a freeze had been imposed on the use of all federal land in Alaska, which is 97 per cent of the whole state, much to the anger of those pushing for development. It was one of the factors holding up the building of a pipeline across Alaska, without which North Slope oil cannot be taken out and marketed. Like everything else, it was oil that had given urgency to the issue.

'Yeah—and that's another thing.' The first man again. 'This is as much my country as any damn native. What right have they got to a free hand-out? That's all they want—something for nothing.'

While the argument went endlessly on, I thought of the terrible poverty I had seen in some of the Eskimo and Indian villages. Of the lack of hospitals and schools—of the fact that the average life expectancy of the 55,000 Alaskan natives, who account for more than one-fifth of the total population, is only thirty-four years, half the average for American citizens as a whole. And of other chilling statistics to do with unsanitary living conditions and unbalanced diets and infant-mortality rates. Yet these were the people who were living in Alaska for centuries before the white man came.

And the battle over ecology that so irritates those with a pioneering drive to tame the wilderness and to get development going. I remembered Art Davidson, dedicated conservationist and one of the team to make the first winter ascent of Mount McKinley, passionately putting his case for preserving the wilderness he loves so much. A few years ago he and those like him were regarded merely as cranks. But attitudes in the rest of the United States were changing. After a century of progress the civilisation that pioneers and industrialists had fought nature for was seen to have led to polluted rivers and smog-laden cities and scarred landscapes. There was a sudden concern for the environment, rising in a groundswell until it became a major political issue—inspired, so the cynics would have it, by the Nixon Administration in order to divert the attention of the young from the war in Vietnam. In most cases it was too late to do much about conservation and there could only be an attempt to reduce pollution. The devastation and plunder of old frontiers was regretted, now that they no longer existed, but nothing could be done to bring them back. The belief in the doctrine of

Manifest Destiny which held that the United States had a God-given right to possess the entire American continent was something that Americans preferred to forget, its sins buried in the past. But then, with the North Slope oil discovery highlighting Alaska, it was found that a frontier did still exist. Here was a land where something could be done about conservation. And the unimaginable happened. Davidson and his Friends of the Earth, together with other conservation organisations, succeeded in winning a court injunction in Washington against the building of the trans-Alaska pipeline until it could be proved that it would not damage the environment. For the first time in American history a multi-billion dollar industrial development was held up in the name of preserving the wilderness.

The conservationists can go too far in the heady triumph of their initial successes. To hear some of them talk—I remembered a long argument with the militant Professor Charles Konigsberg in Art Davidson's home on a remote hillside outside Anchorage—it would seem that mankind should return virtually to the cave. No cars, no packaged goods, nothing that did not come directly from nature. It might be possible for a few people in places like Alaska, but it is no answer for the rest of humanity. And the arrogance of a publication like *Ecotactics*, the Sierra Club's handbook for environmental activists which accuses journalists of being 'fat and irresponsible' and 'effete, impudent snobs' if they do not advocate the ecological cause, curiously seeming to reiterate Spiro Agnew's views, can only harm that cause. Not all who want to develop Alaska are concerned only with self-profit or with tearing up the wilderness for the sake of it. It was a breath of sanity to hear eighty-four-year-old Senator Ernest Gruening, the grand old man of Alaska who did more than anyone else in the struggle to achieve statehood and who was also one of the first of the nation's leaders to make a stand against the Vietnam war before it became fashionable to do so, speak about the need for carefully planned development and at the same time setting aside large areas of land as national parks. One of his dreams had come true with the creation of the Chugach Mountain State Park, where he used to climb and explore in his boyhood.

Complex questions, all of them. Easy to get emotional about, no easy answers. Alaska is in every way a land of contrast. Hunters arguing vehemently for wildlife preservation, yet at the same time supporting the bounty paid for killing wolves which leads to the profitable business of machine-gunning them from a plane. You realise it is only certain kinds of wildlife they want preserved. Eskimos living in animal-skin shacks and still hunting whale and walrus, yet riding around on snowmobiles—the dog sledge is a thing of the past. The town of Anchorage, disappointingly set out in a monotonous grid pattern like any small Midwest town with its flashy neon signs, squat buildings, fall-out shelters, Baptist churches, flag-decked second-hand-car lots, Washeterias, Eats, and Hoc-it-to-Me pawnshop, yet many of the people who live here—nearly half the population of the state—eat only the meat they kill from time to time in the surrounding hills. It is not at all unusual to be asked to dinner in a typically suburban house and find the casserole one is served is made of moose which the housewife has shot herself. And next to the second-hand car lot you may find a second-hand plane lot. More private planes are owned here *per capita* than anywhere else in the world—one to every 100 people—for with few roads and only one major railway this is the only means of getting around the state. One in five Alaskans has a pilot's licence and near the main airport is the only air-traffic control tower in the world devoted entirely to float planes. In the winter they change the floats for skis.

Beneath the surface of modern convenient living still lurks a sense of adventure and danger. At any time in ordinary conversation with an Alaskan about politics or education one is liable to hear a story about a plane crash-landing on a glacier or someone trapped up a tree by a prowling bear or a trapper frozen to death in the woods. The frontier sense of justice still applies. I heard the phrase 'Spenard divorce' several times before it was explained to me by a lawyer. Spenard is a rather tough district of Anchorage, and it simply means that if a woman shoots and kills her husband it is assumed she had a pretty good reason for doing so and the normal sentence is six months suspended if the murder is a first offence. Obviously, it must not become a habit. The sentence for a man is usually no more than a few years, and might also be suspended if he could prove undue provocation or self-defence. The map of Alaska in the office of a conservatively dressed bank manager is full of such names as Dead Horse, Purgatory, Old Woman Cabin, Burnt Paw, Dime Landing, Crooked Creek, Big Hurrah, and Butch Mountain. And at the weekend the manager will likely put on his parka and his fur boots and go to one of those places.

'When people here don't agree about things they'll sure let you know about it.' The late Larry Fanning was publisher of the *Anchorage Daily News*, one of that special breed of independent frontier newspapermen, and his remark summed up the Alaskan attitude. Controversies are there and these courageous, obstinate, friendly, opinionated people aren't afraid to talk about them and don't give a damn who hears. As with other frontiers in other times, it was only the adventurous who came here in the first place and only the most hardy of these who stayed. They are as formidable as the land they have made their home. And behind so many of their attitudes and suspicions are the two great factors about Alaska that are essential to any understanding of this strangely beautiful land. One is its size geographically. The second is its turbulent history. It is an angry story of man's blind greed and stupidity, his inhumanity and cruelty, in which none of the empire-building nations of those times, particularly Russia and the United States, can take much pride. It is told here not to condemn from hind-sight, but to explain some of the very real bitterness that still exists in Alaska today, especially among the Eskimos and Indians. And it is not all darkness. Through the long night of commercial avarice and official indifference are the acts of a few individuals who did show concern and who kept alive a faith in the human spirit.

2

The Great Land

There's a saying among Alaskans that Alaska is always being discovered but is never remembered. It was discovered by Russian fur traders in the mid-eighteenth century, who for a hundred years savagely exploited the vast herds of seals and sea otters off the western shores until whole colonies were wiped out and some species became extinct. It was discovered by two of the great British explorers of the Pacific and the Arctic—Captain James Cook and Sir John Franklin—who named and charted much of the coastline. It was discovered by American whalers and traders, leading to its purchase by the United States from Russia in 1867. The sum paid was $7,200,000—two cents an acre and considered by most Americans at the time to be an act of incomparable folly. Events have shown this biggest real-estate deal in history to be just about the best bargain ever.

Having bought Russian America and renamed it Alaska—the Great Land, in the language of the Aleutian Indians—the Americans promptly forgot about it and allowed it to fall into lawlessness and neglect. It was not discovered again until the great gold rush era of the 1890s by miners pushing westwards from Canada along the Yukon. Then the gold production declined and Alaska sank back into oblivion until the Second World War when it became the turn of the US military to see its potential. Bases were established in Alaska—but not before the Japanese invaded the Aleutians and a small and little publicised war was fought there until they were beaten off.

Finally, Alaska has been discovered by the oil industry, when in

1968 there was located at Prudhoe Bay on the icy plain of the North Slope the biggest oilfield ever to be found in the United States. And this is one discovery that will not be forgotten. What was until then the most impoverished state in the Union, which only achieved statehood in 1959 after years of struggle for recognition, has become potentially one of the most wealthy. That one oilfield alone has been valued at sixty billion dollars* at least, and its oil will be sorely needed as the United States continues to run out of production from existing domestic sources. But the discovery is also of far-reaching global importance. Scientists had for years conjectured that large amounts of crude oil and natural gas might be locked in the fastness of the Arctic region—that only here might there be a chance of finding oilfields comparable in size to those of the Middle East. Now, Prudhoe Bay has shown this to be a fact. The Arctic has become the oil industry's main centre of activity. And leading the race to develop Alaska's resources is a British company—British Petroleum—whose original exploration on the North Slope, was to a large extent responsible for the actual discovery made by Atlantic Richfield and which owns more of the Prudhoe Bay reserves than all the other companies combined—some 60 per cent of the whole field, in fact. It is not unfitting that Prudhoe Bay was so named by an Englishman, Sir John Franklin, who discovered it during his overland journey of 1825–7 to map the northern coastline of Canada and Alaska.

It might seem strange that this most north-western corner of the world, bordering the international date-line which considerately bends to avoid giving Alaska two days at the same time, should have remained in obscurity for so long. Although its great wealth in fur and fisheries has been exploited, these are only a small part of Alaska's immense natural resources. Even apart from oil its mineral potential is enormous, including vast deposits of copper, iron ore, coal, mercury and gold. Alaska is so vast and areas of it so remote that it was beyond the comprehension of most people and certainly of its early political leaders, firstly in St Petersburg and then in Washington.

Alaska's 586,400 square miles are bigger than Texas, California and Montana combined. A map of Alaska superimposed on that of the United States would show the western tip of the Aleutian Islands to touch Los Angeles while the south-eastern boundary would end at Savannah, Georgia. Point Barrow, the most northerly part of Alaska and only 1,250 miles from the North Pole, would come close to Duluth, Minnesota. On a map of Europe, Alaska would reach from London to Copenhagen in the north and from Lisbon to Athens in the south. At

* US billion = thousand million; trillion = million million.

its nearest point to continental United States it is still 600 miles away, separated by the west coast of Canada from the state of Washington. The Russian territory of Siberia, on the other hand, is only fifty-six miles across the Bering Strait from the Seward Peninsula, connected at one time in the geological past by a land bridge. There are no less than three time zones in Alaska, and the difference in winter temperatures, for instance, ranges from ten degrees Fahrenheit above zero in the south to sometimes a hundred below in the north. Alaska is so large that it has to be considered in six main geographical areas, virtually six states within a state, that vary as much from one another as, say, California from Massachusetts.

To the south-east is the Alaskan Panhandle, a narrow strip of rugged mountains and islands between the border of British Columbia and the Pacific Ocean. The mountains rise sheer from the sea to heights of 9,000 feet, crowned with icefields and glaciers, one of which is bigger that the state of Rhode Island. Between them are deep fiords and inland waterways. Massive forests of spruce and hemlock cover the lower slopes, lush in summer when the average rainfall of 150 inches is one of the highest in the United States. This is the most picturesque region of Alaska, bearing a resemblance to Norway and a centre of the salmon-fishing, canning and lumber industries. Both the old Russian capital of Sitka and the present American capital of Juneau are here, accessible like all the towns in the area only by plane or boat. It is perhaps incongruous that fog and bad weather can prevent politicians from leaving or entering their own capital. But in view of one of the wry definitions of the term Panhandle, that it is shaped like the bowl and handle of a frying-pan in which politicians in the years before statehood cooked up dubious deals to manipulate the rest of Alaska, perhaps there is some justice in this.

The coastal mountains, one of the three great mountain ranges of Alaska, arc westwards round the Gulf of Alaska and down into the Kenai Peninsula, reappearing from the sea as Kodiak Island at the southern tip. This is the south-central region, extending inland to the mighty Alaska Range which cuts 600 miles and in places 150 miles wide across central Alaska and includes the highest mountain in North America, the 20,320-foot Mount McKinley, among its peaks. It is the most populated region of Alaska. Along the coast are deep-water inlets, relatively ice-free in winter, which serve as natural harbours. Prince William Sound is one, with its seaport of Valdez, and on the other side of the Kenai Peninsula is Cook Inlet, leading to the city of Anchorage, Alaska's commercial centre, with a population of about 125,000, nearly half that of the entire state. The site was named by Captain Cook who anchored here in 1778 before sailing southwards into the Pacific to meet his death at the hands of the Sandwich Islanders. Inland between the two mountain ranges are expansive valleys which contain the best farmlands in Alaska and are plentiful with game and wildlife. The region is rich in copper, mined particularly along the valley of the Copper River which flows into Prince William Sound, and gold-mining is also carried out on a small scale.

The mountains of the Alaska Range are snow-covered throughout the year and cut by huge glaciers, yet they can be crossed more easily through the major river valleys than the Coast Range. Merging with the coastal mountains at their eastern end, they fork northwards and sweep down again to form the Alaska Peninsula and the long chain of the Aleutian Islands, curving like the vertebrae of a spine. Together with Bristol Bay they make up the south-west region. This is an area of contrast, extending 2,000 miles from the forested hills and fish-filled lakes of the peninsula to the barren, volcanic Aleutians which reach across the Pacific to within a thousand miles of Japan. The population largely comprises Aleuts and Eskimos, who suffered savage repression at the hands of the early Russian fur traders, and it is here, particularly among the Pribilof Islands, that are located the main breeding grounds of the fur seal and the sea otter. Both species were almost exterminated by wanton killing earlier this century, but they are now protected by conservation laws. The fisheries resources of Bristol Bay, especially salmon, are the most important in Alaska. For most of the year there is no commercial transportation over large parts of this sparsely populated region. The seas around its coast are turbulent and storm-tossed as cyclonic disturbances pass from east to west across the Pacific, for this is where the winds of the northern hemisphere are born. It is a land of fog and rain and winter storms, but the climate is surprisingly mild because of the warmth of the Japan current. There are no trees on the rugged Aleutian Islands, but during the short summer the grass hangs thickly down from the cliffs and even wild orchids grow on the island of Unalaska.

Interior or Central Alaska is a vast area of rolling forest range between the Alaska Range and the Brooks Range to the north, the third of Alaska's main mountain ranges. It reaches 1,000 miles westwards from the Canadian border, more than 300 miles wide, and to many this is the real Alaska. Across here flows the great Yukon River from British Columbia to the Bering Sea—the third largest river in North America after the Mississippi and the Mackenzie. If the Alaska Range is the backbone of Alaska, then the Yukon is its main artery, which during and following the gold rush was the main means of transportation in the interior and on to the Klondike in Yukon Territory. Indian villages and trading posts are scattered along its banks, together with the derelict remains of old mining camps. On the Tanana River, the Yukon's largest tributary, is Fairbanks, second-biggest city in Alaska and located almost in its dead centre. Outside Fairbanks, with its 36,000 population, the interior is a virgin frontier, inhabited by many species of wildlife, including moose, caribou, bear, sheep and hordes of migrating birds, and occupied almost entirely by Indians of the Athabascan tribes. The winters are long and cold, with temperatures falling to minus twenty, but in the short warm summer they can rise to as high as ninety-five degrees. Much of the subsoil, even in summer, is permanently frozen to depths of several hundred feet—the permafrost which is causing the oil companies so much trouble.

The western coast of Alaska along the frigid waters of the Bering Sea consists largely of the Seward Peninsula and the huge delta of the Yukon and Kushokwim Rivers. It is a remote region occupied mostly by Eskimos, but at the turn of the century the gold-rush settlement at Nome on the Seward Peninsula was known throughout the world. What was once a tent city is now a small community of some 2,500. Following the gold rush, reindeer were introduced here to help the Eskimos exist in the harsh environment.

One-third of Alaska lies above the Arctic Circle. Northwards of the little-known and -explored Brooks Range is the true Arctic—the North Slope—a vast plain of permanently frozen tundra 600 miles long and up to 200 miles wide, sloping down from the foothills to the Arctic Ocean. It is featureless except for the rivers and streams which drain into the polar seas, the Colville River halfway across being the largest. For most of the year the coast is ice-locked and cut off from any form of transportation except by plane. Land and sea merge into one—a white flat wilderness of ice and snow. This is the land of the midnight sun. During the two months of summer it never sets, but although the temperature rises to little above freezing, the ground becomes covered with grass and wild flowers, a marshland laced by thousands of streams and lakes which are important breeding areas for wildfowl. Only the top few inches of soil melts, however, Below, the ground for up to 2,000 feet remains frozen. During the rest of the year the temperature is always below zero. With one mile of wind per hour equal to one degree of frost—known as the chill factor—a not uncommon blizzard of sixty miles an hour can reduce the temperature to well below minus 100 degrees Fahrenheit. For two months of winter the sun never gets above the horizon and there is perpetual darkness. A large proportion of Alaska's 25,000 Eskimo population live in this region, centred on the villages of Barrow, Kotzebue and Point Hope, with hunting and fishing still a major means of subsistence. Caribou, wolf and Arctic fox range the North Slope, grizzly bears prowl the foothills in summer, while far out among the ice-floes of the Arctic Ocean are polar bear and seal. Whales and walrus are hunted in the spring from frail craft powered by outboard motors—a concession, like the snowmobiles on land, to technological progress. Barrow is the most northern point of Alaska, from where many of the polar expeditions have set forth across the ice to the North Pole.

These six regions of Alaska are among the very few areas in the world that still remain to be settled. The main means of transportation is by plane. Indeed, in many places there is no other way. The only major railroad in the state is the 470-mile line from Seward on the Kenai Peninsula to Fairbanks, via Anchorage. The service runs twice a week and with charming informality the train will stop on the way to pick up hunters and travellers who flag it down between stations. There are relatively few roads in Alaska—a fact that is a major headache to the oil companies. Best known is the 1,520-mile Alaska Highway from Dawson Creek in British Columbia to Fairbanks, although most of this is in Canadian territory. Other highways connect the main cities of the south-central and interior regions—Valdez to Fairbanks, Seward and Homer on the Kenai Peninsula to Anchorage, Anchorage to Tok Junction and Eagle near the Canadian border, and a road being built from Anchorage to Fairbanks—but elsewhere they are few and far between. The only roads at all north of Fairbanks are to Circle City, on the Yukon, and to Livengood. The construction of a temporary road from Livengood up to the North Slope to take equipment to the drilling sites caused a storm of controversy, as will be seen later.

It is the airplane that has opened up Alaska, taking the place of train and bus as the regular transportation between communities. Alaskans on average log more airline trips than anyone else in the world. More than 300 communities, some of them no more than small Eskimo settlements, are served by a regular air service. This was the country pioneered by the famed bush pilots—men like Neil Wien, whose pilot's licence was signed by Orville Wright, Carl Ben Eielson, a Fairbanks schoolteacher who flew the first flight across the Arctic Ocean in 1929 before crashing to his death the following year off Siberia, and Bob Reeve who was first of the glacier pilots. The exploits of these and others like them are legendary in the northland. Their successors are still keeping up the tradition—fliers such as Don Sheldon and Bud Helmericks—but the days of the true bush pilot are fast passing with the advent of the jet age. Officially, they are already over, for the

Federal Aviation Agency has changed the bush-pilot certificate to air-taxi operator. But flying in Alaska, even with the most modern aircraft and all the latest aids, is no easy matter. Air-freight services received a great impetus when oil operations started up on the North Slope and much of the equipment had to be flown in, but a higher than average number of planes crash in blizzards and storms and white-outs which can occur so suddenly with little warning.

Few of those who came to exploit Alaska's wealth—its gold, fisheries, fur and minerals—intended to stay. Whether or not they achieved the riches they sought, the land usually beat them in the end. Only the most determined stayed, those who had a genuine feeling for the wilderness, so that even today the population of Alaska is only 280,000—about one person per two square miles—and three-quarters of these live in the three main cities of Anchorage, Fairbanks and Juneau. As a comparison even the relatively thinly populated state of Texas has thirty-six people to one square mile, while the average population of England is 826 per square mile. About one-fifth of Alaskans are Eskimos and Indians, the original inhabitants who were here long before the various 'discoveries' were being made by the explorers of other countries.

The picture most outsiders have of Alaska is a wilderness of snow and ice with miners panning for gold in lonely creeks and a gaudy saloon at the end of the trail. The land of Dan McGrew and the spell of the Yukon. Jumbo-jets and modern buildings in Anchorage and Fairbanks might seem to spoil this image, but outside the cities it is still a pioneering country. Men still go out to dig and pan for gold, although it is no longer a major industry. It is still possible in the wild to find valleys and mountains which are unmarked on maps. It only needs twenty-five people, of whom nineteen must have lived in Alaska for at least three months, to get together to form a township of their own and call it anything they like. Many of the new Alaskans came after the Second World War, turning their backs on security and comfort to seek the freedom of open spaces. Others came earlier. People like Elmer Keturi and his wife Hilda. They live in Fairbanks now. Elmer is a tall, leathery sixty-five-year-old, retired after an adventurous life of gold-mining, breaking new trails to haul freight, working in construction and as a mechanic. His wife taught school for Eskimo children for many years.

'I was brought up on a farm in New York State,' Elmer says. 'I had an uncle who used to commute every year from Detroit to Alaska. This was in the 1920s. Every spring he'd take a train to Nenana, just near Fairbanks here, then walk 400 miles cross-country to a little settlement called Flat on the other side of the Kuskokwim Mountains. He'd work there all summer, then walk 400 miles back in the fall. Early in 1929 he visited us and asked if I'd like to go with him. I hated farming and I said, hell, yes. It took us seventeen days to walk to Flat. I was twenty-two then and he was close on fifty. Flat was a boom town at one time. They must have taken millions in gold out of there. But after World War II it started to go down and, well, it's a ghost town now. But there were still three or four hundred people when we were there. Later on, most of them would come out for the winter. That was when the planes started hauling passengers and charging only a fifty-dollar fare. But it was $350 one way in 1929. People stayed in those winters—they couldn't afford it, you see. It was either that or walking out or coming by dog-sled over the trail. They hauled the mail by dogs in those days and that was the trail my uncle and I used to follow when we went in.

'I stayed for two summers working on a gold dredge. We'd work twelve hours a day, seven days a week, for three dollars a day. It was hard, I can tell you. Then I went out and got married and we both came back in 1934. Been here ever since. I worked some more on a dredge and then in 1936 four of us went into partnership to dig for gold on our own. We took over a mine a fellow went broke on—well, he didn't exactly go broke but spent all his money in the winter on girls down the line and the bank foreclosed on him. We didn't have much luck the first summer but then we got some better equipment and made good money.

The mine was at Moore Creek, forty miles north of Flat. In 1937 the Keturis, who had a one-year-old son by then, decided to spend the winter there with another couple.

'It was getting close to Christmas and we figured we'd go to Flat for a dance being held there on Christmas Day. We just had this little caterpillar gas tractor, so we fixed up a kind of tent on a sled for the girls with an oil heater in it and the girls cooked up a mulligan stew in a big cast-iron kettle. It was all we had to last us until we got to Flat. We had to break trail all the way. I was driving and Tony stood out in front. We kept having to cut down trees to stop the branches swinging back and hitting the tent. Soon after we got going we hit a log and the sled lurched and the stew got upset all over the floor. Luckily we'd put down a clean sheet of iron on the floor to stop the brush breaking through so the girls managed to save the meat, but they lost the liquid. It was getting time to eat and I was pretty hungry, so they had to melt down some snow and re-cook it. Sure was a good stew, made of moose, and they never told Tony and me a thing about it being spilled until afterwards. It took us two days to make the trip—about a mile an hour.

We got to the dance and there weren't many women, so the girls had to dance all night until six the next morning. Then we headed back to our cabin.

'In 1942 the government made us quit mining gold because it wasn't helping the war effort. The four of us with our wives came back to Fairbanks and we had these tractors, big D.8s and Internationals, and had to find something to do with them. So we got a government contract hauling freight to where they were building airfields. On one of those trips we had six D.8s hauling 600 tons of runway mats on bob-sleds—go-devils, we called them. Two tractors had to go ahead to break trail because there was five feet or more of snow. We got to the Yukon and drilled down and found there was four feet of ice so the tractors would go over. But then we came to a place where the ice wasn't so thick and darned if the lead cat didn't go straight through. Just disappeared. The driver, a Finnish guy called George, he went down with it. There was a cable attached from the cat to a wanigan—that's a shack built on timber sleds to use as living quarters when you're travelling—and it was jerked right to the edge of the hole. Luckily there was only twenty feet of water or the wanigan would have gone down on top of George. Anyway, there he was in the water looking up at the hole in the ice above him and it was already getting smaller as more ice began to fill it in. In the ordinary way he'd only have lasted a few seconds in that temperature, around minus fifty it was, but he had a lot of heavy wool clothes on and big Indian gloves. He came up like a cork. He was bald, I remember, but had a fringe of blond hair and Nick the other driver grabbed hold of this and pulled him out. He was shouting, "T'ain't safe, t'aint safe!" Luckily the water hadn't got all the way through his clothes. It was two miles back to the camp we'd set up and he had to stand on the drawbar of the other tractor all the way and, by golly, he was just about a frozen icicle by the time he got there. It was thirty below and the wind was blowing and I don't know how he stood it. We put dry clothes on him and he never even caught a cold.

'Well, in 1945 they let us go gold-mining again and we started up there in the Arctic near Wiseman, east of Bettles field. We got our tractors back and made some more sleds for hauling in our supplies. We'd bought a mine from an old-timer who'd been there twenty years and took him in as a partner. We had to break trail all the way from Livengood. I was in the lead cat because I'd flown over the route beforehand and had memorised the landmarks.'

This trail was the first ever to be broken northwards from Livengood. When the Ice Road was made in 1968 for hauling oil equipment up to the North Slope the trail was still plainly visible and the new road followed it for most of the way.

'We left Livengood in mid-February 1945. There were five of us in the team, including the old-timer. We dropped down into Hess Creek for quite a way. I wanted to hit Stevens Village right on the nose because we had barrels of oil there that we'd brought up by river the year before. Then it started to snow and, man, I've never seen a snow-storm like it. I couldn't see ten feet, let alone any of the landmarks. The only way I could see I was going north was by compass and even then I had to leave the tractor and walk some way ahead because all that steel made the compass go crazy. The storm lasted for days. In heavy timber we were lucky to make two miles a day, but over scrub and spruce we could make about three miles an hour. Then we came to a long creek and I decided it was bound to lead to the Yukon flats. Sure enough, it did, but I didn't know if we were above or below Stevens Village. I had a hunch and turned left and then after a ways I found some snow-shoe tracks. So I got off the tractor and followed them and came to a little cabin where there was an old trapper cutting wood. He told me we were forty miles up from Stevens Village. So we carried on along the bank of the Yukon and ran into the most godawful willows, so thick you could hardly get through even in a D.8. It was terrible, worse than trees, just like going through jungle, they were so springy and hard to break. Then we had to cross the Yukon and it meant drilling holes to see how much ice was there. We needed at least twenty inches to be safe with a twenty-three-ton tractor and most of it was only twelve inches. It took us a long time to find a good place to cross. But we weren't taking any chances after our last experience on the Yukon, so we put the cats into gear and let them go across by themselves, steering them with ropes.

'We got to Stevens all right and picked up our oil and machine parts. North of there we ran into hot springs, boiling and steaming away, and there was hardly room to get through. It took us until May to haul everything up to Wiseman. It's only 200 miles from Livengood by air, but the trail was about 400 miles long. We found gold there all right but the place was a mass of big boulders and frozen permafrost. We did some surface mining for about five years but it hardly even paid for itself. We had our families with us and built our cabins on skids so we could move camp when we wanted. The first thing we built was an airstrip in case of emergencies—and it was just as well because that first year we were there our son, who was nine then, fell over and broke his arm. We put it in splints, but it was still seven days before a plane could land and take him to hospital. We had to quit mining in the end and sell up the machinery to help pay our debts. I went to work as a mechanic and Hilda taught school.'

One of Elmer Keturi's partners at one time was Joe Fejes, who also now lives in Fairbanks. His wife Claire is one of Alaska's best-known artists.

'In the days before tractors the old gold prospectors sometimes used dogs and horses for freighting supplies, but mostly had to rely on their own manpower. In the winter they'd build boats, whipsawing their own lumber, then put these on toboggan-like sleds and pull them by ropes fitted round their shoulders. Necking they called it. They'd go for miles over frozen ground before setting up camp. The way they used to mine was like this. They'd build fires to thaw out a hollow in the ground, then dig out the thawed earth with a shovel and bucket, and repeat the process, doing deeper all the time. It was very slow, but they'd go down as far as 100 feet sometimes. No shoring was necessary because the ground was so frozen. When they reached bedrock they'd start panning the stuff from the bottom of the hole. If there was something that looked like paydirt they'd start tunnelling, using the same process of thawing by fire and shovelling the earth out and winching it up in a bucket. By this time they'd usually have a partner or would have hired someone to help. Some would stick it out by themselves—even today a few individual prospectors are working claims—but it was risky. Most of the ones who didn't come back had gone out and stayed by themselves. The wilderness is an awful dangerous place. You can't fight Mother Nature because she just takes you right down and, boy, you're finished. For instance, you get a deep snow. The moose and the rabbits, they'll scrabble down and chew off any tender shoots that are still coming out of the ground. When the snow goes down you've got hard spikes sticking up and if you fall down they'll impale you. When you walk over the ground in summer it's like sharp steel spikes sticking up. You'd bleed to death before getting any help.

'Then there's bears. They're fast—they'll get you before you can get to a tree. Or maybe there isn't a tree around for you to climb. I came face to face with one once and the biggest thing around was a fireweed—four feet tall. If he'd charged, I wouldn't have had a chance. I picked up a clump of dirt and threw it at him and shouted and he just walked off. He was hungry, but not for me, I guess. An ordinary black bear won't bother you unless he's in pain, like with a toothache, or it's a female with cubs.'

Joe first came to Alaska in 1943 with the Army after being drafted in New York City. By the time he came out of the Army he had got the gold bug.

'I had it bad. And the only way to get rid of it is to go out and do it. If you make money, you're lucky. If you don't, at least you've got it out of your system and if you're smart you'll give it up. The price of a tractor then, fully equipped, was around $17,000. It would be $75,000 today. The small miner is mostly out of the picture now. What people do today, if they find something good they try to peddle it to the big mining firms.'

His first experience of mining for gold, as distinct from the panning he'd done while in the Army, was traumatic. He had returned to Fairbanks after the war with his wife and found a site sixty-five miles north at Nome Creek. A mining company had been looking there, but found nothing, so Joe was able to buy up some of their equipment with his savings. But even with modern equipment the wilderness is a dangerous place, as Joe was to discover.

'We tried to get the tractor back to Fairbanks in December so we could load up with supplies to take back the following spring, but there was too much snow. In March I met another guy who had a tractor of his own up there. His plan was for both of us to fly up and then drive the two tractors back. He said he had food at his camp. So I agreed. We flew up and than walked on snow-shoes to his camp, carrying batteries on our backs for the tractors. It was five in the evening when we got there and I found the food he'd talked about was just some hot-cake batter and a few cans of spaghetti and peaches. It was thirty below at night but just about thawing during the day. That was bad because it meant new ice was being formed all the time.

'We put the batteries in his tractor, then we found the temperature gauge was in the red. What had happened, in the fall some hunters had shot a moose and wanted to bring it back across the valley so they'd borrowed the tractor—and as a big favour they drained the water from the radiator, not knowing it had been winterised. I got the biggest container I could find, which was a two-pound coffee can, to fetch water for the radiator. It took about forty gallons, so that meant a lot of trips to the stream. I chopped a hole in the ice and started. On one trip, when it should have been about half full, I noticed water trickling down the track and found the plug at the bottom of the radiator was still open. The water was coming out as fast as I put it in. Anyway, we finally got it filled and tried to start it. It wouldn't start. Either it meant a long walk back to Fairbanks or we had to make it start. There was a construction camp a few miles away where we found a two-and-a-half-horsepower air-cooled engine. We lugged this across and used it to start the generator pumping juice into the batteries.

'It was around ten at night by the time we got it going. We went to bed, leaving it running. After a few hours I woke up to hear—nothing. The engine had broken a rod. So the batteries weren't charged and we were in the same situation as before. Next morning we managed to find another engine—six horsepower this time. It was too heavy to carry from the construction camp. But in the cookhouse we found a big pan that the cooks used to bake bread in. We bent the front end up like a sled, put a rope on it, and pulled the engine over to our camp. We'd arrived there on a Sunday. It was the following Saturday by the time we got the tractor going. All this time we'd been living on spaghetti and peaches. Claire was in Fairbanks and getting very worried, of course but there was no way of contacting her.'

The families of the two men had in fact been in touch with the authorities. Soon after they had got the tractor started, and were on their way to the place where Joe had left his own tractor parked the year before, a plane came over and dropped a message asking if they were all right. They stamped out the word 'Okay' in the snow and the plane flew off.

'By the Saturday evening we were heading for the highway. The road out of the camp was snowed in, about five feet. The guy I was with and who was driving lowered the blade, but the snow was too thick and we kept slipping back. It was five miles to the highway and by midnight we hadn't gone more than a mile. It was twenty below and there was no heating in the tractor. It was noon the next day by the time we got to where I'd left my tractor back in December. We had to chop the ice from it, melt some snow to fill the radiator and fill up. Then he took off and I was following. We had to decide either to take the road or the winter trail through the woods. I was all for the trail rather than fighting the drifts on the road when you spend all your time just pushing snow. but he said he didn't know the winter trail. You could see it in fact—when once a tractor has gone through it leaves ruts in the ground which fill with water in summer and willow saplings grow. But he said we should stay on the road and I figured he was the boss, so okay.

'We started out. I'd never thought much about glaciers before, but they were something, I can tell you. Hills will absorb a lot of moisture, especially if covered with snow, and when it starts to thaw it causes temporary glaciers which you find on all hillside roads. A barrier of solid ice and snow. It looks like the line of the hill, but you know there's a road underneath somewhere. When you're on top of this you're in a bad situation because although the tractor has cleats on its tracks, on top of the ice it's like a sled because there's nothing to grip on to. When we hit the first glacier I found my tractor sliding sideways. I was ready to jump in case it went over the hill, but it stopped just in time. His tractor also slid sideways until it was hanging over the hill but the heavy blade in front kept it from going right over. That shook us both up. We dug under one of the tracks of each tractor and managed to get them off the glacier. I had a can of corned beef I'd brought with me from Fairbanks for emergency. I figured this was an emergency—we had nothing else with us to eat. We though we'd be in town by Monday. It was now Monday night and we'd only just started. He had on an Air Force parka, but I had just a parka liner and thick socks and snowshoes. I had thought we'd be working most of the time and foolishly hadn't come prepared. It was cold, man—minus twenty-five. This was now eight days after we'd left Fairbanks. We drained some oil and built a fire. It kept going out, then we'd wake up and put on some more oil. It was a terrible night.

'By 4.30 the next morning we were digging our way through another glacier on the road. Then we began making good time, about four miles an hour, until we came to a stream called Goose Creek. Around the corner the road went uphill and we were sliding all over the place. In the middle of the road there was another glacier with a mound in the middle of it, as high as the width of the road. He was heading straight for it and I could see he wanted to climb over it. I threw my clutch and stopped. He stopped too. I said it was too dangerous to carry on and suggested taking the valley trail. He was impatient to get on—it was his birthday the following day and he wanted to get to Fairbanks in time for it. He insisted on going on, even though there was a drop of sixty feet over the side of the road. I wasn't going that way, so we unhooked the cable we'd had strung between us. He got back on his tractor, under a little home-made canopy he'd built to give himself some protection from the wind when driving, and set off. He climbed on the glacier, then started slipping. The tractor went off the road just like a bullet. He looked back at me—you could see he was thinking, Well, this is it, Joe, goodbye. The tractor rolled over, went down the steep drop, and flipped on to its back. It was over in seconds. There was nothing that could have been done to stop it. When the noise stopped I went down to look for him. I thought he might be between the tracks, but he'd been thrown clear. I jumped down the hillside, thinking he might be not too bad, landing in that soft snow. He was kind of twisted up, one arm stretched out maybe five feet long, and his fur hat pushed down over his face. I pulled the hat back to look at him and got his brains in my hand. He'd been crushed to death. Man, then I went crazy. That was the weirdest experience I ever went through. It happened at noon. I had to hike in on foot. The glacier got steeper further on—we'd never have got the tractors across. I had snow-shoes on, but they'd frozen to my feet and I couldn't get them off. I had to climb across on my hands and knees, hanging on to bits of brush sticking through the ice.'

It was late that night when Joe, numb from the cold and still in a state of shock, had staggered for miles through the snow and come to a trapper's cabin. He telephoned for the Highway Patrol to send a helicopter to pick up the body. He could have stayed in the cabin until help arrived.

'But I just wanted to get on. I wasn't even hungry. I figured in a few hours I'd be in Chatanika. It was getting very cold and the moon came up and I'm wiggling my toes because I know they're going to freeze. And I'm crunching along in the snow and start talking to myself and think I'm going crazy. I'm looking for a windfall where I can build a fire, but there isn't any place. Usually you can find them everywhere, a place where a tree has fallen over and you can crawl under shelter, but there wasn't anything here, just when I needed it. It was too dangerous to build a fire in the open because I might fall asleep beside it and the fire would go out and I'd freeze to death. So I had to keep going. About five miles out of Chatanika I saw two figures coming towards me. They weren't wearing snow-shoes and I could see first one then the other fall into the deep snow off the road and the other would help him out. They were two natives, both drunk. One was a well-known dog-musher. They had a bottle of whisky and they gave me a drink and, man, then I thought I really wouldn't make it. With no food inside me it went right to my head and I got twice as cold. I was staggering and breaking out in a cold sweat. Then round a bend in the road I came to a little tent and in it was a friend of mine, an electrician working outside as a linesman. He cut off my snow-shoes. My feet were blue. He started rubbing them. They were okay. I'd walked all day and a night, but I couldn't get to sleep. The next day the Highway Patrol came along in a tractor and drove me to Chatanika where I got a lift back to Fairbanks.'

The frontier town, with its log cabins, sidewalk saloon and dirt main street, still exists in Alaska. Like the village of Talkeetna, 160 miles north of Anchorage and only recently connected to the outside world by road, since it is near the highway being built to Fairbanks. Except that, unlike the frontier towns of old, most of its 120 inhabitants live on welfare and social-security payments because there is little work in the area. Some of the men go out into the hills in summer to dig for gold. And there is always the meat provided by trophy-hunters during the shooting season, for it is a standing rule in Alaska if you kill an animal you must bring the meat out and if you don't want it yourself you must give it to someone who does. The small hotel in Talkeetna that hunters use is owned by Alice Powell. When she came to Alaska in 1947 at the age of forty, Anchorage had only just set up a Department of Health to consider such matters as sewerage disposal and control of water supplies. She became one of its first public-health inspectors covering a vast area of south-central Alaska. One of her jobs was to inspect the sanitation in restaurants and rooming houses. But she soon found out that many of these were a cover for brothels, or 'sporting houses', as they were called. They were officially illegal in the state, but there were few police to enforce the law, even had they felt so inclined. The Department of Health had no rule to say they could not exist, as long as they conformed to sanitation regulations.

'I remember one rooming house in a back alley run by a woman who had made a good deal of money in her day,' Alice says. 'She operated what she called the chili parlour. Actually I found there wasn't any food at all, but it had a counter and pots and pans and all that. The menu hung on the wall, written by hand, and it offered things like Mexican dish, or Chinese dish, or Southern dish. A customer would get the kind of dish he ordered—only it was a girl, of course. I certainly couldn't find anything wrong with the sanitation in that place—everything was spotless because it was never used as a restaurant anyway.'

Alice had the authority to close such places, not because of their illegality but if they had bad sanitation. It was for this reason that she acquired the nickname 'Evil Alice' from the resentful madams in charge. When she visited some backwoods village the word would go round, 'Here comes Evil Alice'. Inevitably her job developed into one of looking after the health of the girls in these places.

'I used to be threatened sometimes, but no one ever attacked me. It was easier for a woman. A man would have been in real trouble. As a matter of fact, they couldn't find a man to do it.'

It is people such as Joe Fejes and Alice Powell and Elmer Keturi who were the modern pioneers of Alaska, following a tradition that went back to the days when the Yankee seafarers first visited this northern land and some of them liked what they saw and stayed. But many centuries before them there were other people from Asia who at some time before the dawn of recorded history were the true discoverers and settlers of the American continent. It is most likely that the original migrations were made into Alaska by way of the Bering Strait from Siberia and to the Aleutians from Kamchatka and the string of islands leading northwards from Japan. Some of these early migrants pushed on to settle North and South America—others remained in Alaska. Recent archeological excavations have shown a remarkable similarity between the ancient carvings of Peruvian and Alaskan Indians. Those who remained in Alaska formed into four main groups: the Tlingit and Haida Indians in the south-east, the Athabascan Indians along the south coast and the interior, the Eskimos on the coast of the Arctic Ocean and the Bering Sea, and the Aleuts, related to the Eskimos but with a different language and customs, in the Aleutian Islands and parts of the Alaska Peninsula.

For many centuries these northerners lived in isolation while the world was gradually being rediscovered by peoples of the Mediterranean and European countries. The great voyages of exploration in the sixteenth and seventeenth centuries had charted the shape of most of the world's seas and continents, but one area remained missing. This was the North Pacific. Northwards of California and the vast area of Siberia there was a blank on world maps. It was not even known whether the converging continents of Asia and America were joined, or separated by water, or even by an entirely undiscovered subcontinent. This continued to remain a mystery until well into the eighteenth century, when it was solved by the Russians, making the last great land discovery on earth and completing the outlines of the major continents.

3

Under the Russian Bear

The first recorded discovery of Alaska was made in 1741 by a Danish sea captain, Vitus Bering, in the service of the Imperial Russian Navy. It is more to the point perhaps that this was when the natives of Alaska first discovered the white man and it was hardly for them an agreeable experience. Civilisation brought little but misery to the Eskimo and Indian, the effects of which are apparent to this day. The relatively peaceful, extrovert Aleuts were the ones to suffer the most, bearing as they did the brunt of the invasion. In the wild scramble for the valuable fur of seals and sea otters which were found in such profusion in the seas off the Aleutian Islands they were rounded up in their thousands and forced to hunt for these animals while their women were kept as hostages and for pleasure. Torture, massacre and rape followed the Russian freebooters from island to island along the Aleutian chain. They were possibly no worse than other European and American colonists, but the fact is that within forty years a population of some 25,000 natives had been decimated to little more than one-third of its original figure. One Russian captain with an obviously enquiring turn of mind lined up a dozen Aleuts in a row behind one another to see how many bodies a musket-ball would penetrate. The answer was nine, which says something for the power of those early muskets.

The origins of this invasion go back to the sixteenth century when with a determination seldom equalled in history Russian pioneers conquered and occupied the vast unknown wilderness of Siberia. Many of them were outlaws, army deserters and runaways from the forced-labour camps which even then had been established in this bleak region as places of exile for political prisoners unwanted by the governing regime in Moscow. Together with traders and adventurers, what they all sought were the rich furs of sable, ermine and bear that had been so valued by their Mongol overlords and were now in great demand in China. Following then came the Cossacks and regular army troops to build a chain of forts across the new territory and empowered with the task of collecting taxes and tribute from the native tribes. For where there were good fur grounds the natives would have the means with which to pay taxes.

Conservation was an unknown word in those days. As areas became depleted of animals, so the hunters had to move further afield. They penetrated Mongolia and Manchuria to the borderland of China, and the time came when they reached the Pacific shores of eastern Siberia and the Kamchatka Peninsula, an unmapped region of snow-capped mountains and dense forests and marshland. Here for the first time they came upon a new animal. Frisking in the surf just off the coast were herds of sea otter, whose pelt was found to be even richer and more beautiful than sable. For the moment, however, there was little the hunters could do except watch, for the animals would appear one day and then disappear the next, seeming almost to challenge the hunters to come and find them. It was highly frustrating, for the skins of the few that were caught were greatly prized by the merchants in Peking. One pelt could command as much as a clerk's yearly income. Where did the sea otters come from? The natives vaguely pointed to the east and said they came from a land which lay in that direction. That land was Alaska.

It was not only greed for furs which brought the Russians into Siberia. In 1689 Peter the Great had come to power and after years of barbaric Mongol rule and then, squabbling amongst the Russian overlords which had kept the country in a backward state, he set about creating a new civilisation in Russia. He turned to the West for learning and knowledge and built an entirely new capital, St Petersburg, on the shores of the Baltic, encouraging European scholars and scientists to come to the various institutions he founded. One such institution was Russia's first navy and one of its officers was Vitus Bering, born of Danish parents in 1681 and commissioned when he was twenty-three. Peter was a man of great curiosity and with a great respect for science. On his visits to Europe he found that the question being most hotly debated in scientific circles was whether the continents of Asia and America were joined in the north. He became obsessed by the need to find out the answer for himself and in 1724 ordered an expedition to cross overland to the Siberian coast and there to construct a ship to make a voyage of exploration eastwards. As expedition leader he chose Bering, then forty-three. He wrote Bering's instructions himself, charging him to find and chart the American coast. It was Peter's last official act, for he died a few days later on January 28, 1725. By then the advance party of Bering's expedition had already left St Petersburg. Peter was succeeded by his widow, Catherine I, who followed his plan through. Bering and his lieutenant, Alexei Chirikov, set out with the main expedition in February.

Bering's problems were formidable. All the equipment, except for lumber which was obtained locally, had to be hauled across 6,000 miles of wild country to the shores of the Sea of Okhotsk. This alone took two years. At Okhotsk two boats had to be built to take the expedition 600 miles across the bay to Kamchatka. There a larger ship, named the *St Gabriel*, was built and in July 1728 Bering set sail northwards on his mission. He was back in less than two months, having discovered a large island which he named St Lawrence, and which is now part of Alaska, and decided to his own satisfaction that the continents were not joined by land. But he did not have conclusive proof. He had not sighted the mainland, even though he had sailed through the strait which now bears his name and from where on a clear day both the Siberian and Alaskan coasts are visible. Bering had found the area shrounded in fog and rain and did not wait for the weather to clear. On his return to St Petersburg he was severely criticised by the scientists for not remaining until he had completed the task.

A second expedition was ordered, again with Bering in command, but it was not until 1738 that he was able seriously to get down to the task of organising and building the ships required. Years of harassment and political intrigue had held up the project and made Bering a tired and prematurely aged man. The second expedition was far grander than the first. It was made up of 900 people, including a number of scientists and one in particular, the German-born naturalist Georg Wilhelm Steller, whose ultimate contribution was perhaps the greatest of all. Two ships were built, the *St Peter*, with Bering in command, and the *St Paul*, under Chirikov, and then set sail from Kamchatka on a south-easterly course on June 4, 1741. They were squat two-masters, only eighty feet long, with a twenty-two-foot beam and nine-foot draught. It was intended that they should both stay together, but things went wrong from the very start. After a few days they became separated in the fog which seemed to be such a permanent feature of those waters. Some time was vainly spent in trying to find one another, then each decided to go it alone. The ships never met up again—and neither did Bering and Chirikov.

Chirikov was the first to sight the American continent. On July 15 he came to the mountainous coast of what is now Cape Addington in south-eastern Alaska. There followed one of those mysteries of the sea which have never been solved. Three days later, in an effort to find a suitable anchorage and to obtain water, a longboat with eleven armed men was sent ashore. They rounded the point of an island and were never seen again. The signals sent up by the *St Paul* remained unanswered. Five days later the one remaining boat with four men on board was sent after them. This too disappeared and never returned. After waiting for several days but unable to come closer to the shore, Chirikov became convinced that some disaster had befallen the men, possibly at the hands of natives who had been glimpsed in canoes. With water running low, he had to return to Kamchatka as quickly as possible. He brought with him news of the first discovery of northwestern America, but that was all. None of those who returned had even set foot on the new land. It was a particularly barren voyage.

Bering's course took him parallel with the Aleutians, although the deep water gave no hint of land. Eventually on July 16, one day later than Chirikov, he had his first glimpse of the American continent. The clouds suddenly lifted and, to the awe and excitement of all those on board, there, towering into the sky, was a massive snow-capped mountain range with a central peak 18,000 feet high. They had come to the south coast of Alaska, just eastwards of what is now Prince William Sound. Bering named the highest mountain St Elias. A landing on Kayak Island, mainly to obtain water, revealed signs of native habitation. Steller and the other scientists on the *St Peter* were eager to stay to continue their observations, but Bering decided the time had come to return. He had achieved the main object of his mission. But primarily it was because many of his crew were suffering from scurvy and colder weather was beginning to set in. Any delay would mean spending the winter on that unknown and inhospitable coast. Many historians have criticised Bering of weakness and irresponsibility in his attitude to the whole expedition. It was said that on first sighting the St Elias range he merely shrugged his shoulders and was indifferent to the congratulations of the crew. But he was then sixty years old, ill himself from scurvy, and had also been subjected to years of hostility and jealousy in St Petersburg. It is likely that he never really wanted to undertake the expedition in the first place—he certainly had no interest in science. And now it had taken up most of his adult life. Considering all this, it is remarkable that he achieved what he did. As to his attitude towards Steller and the other scientists who wanted to spend as much time as possible examining this exciting new land, it is almost as if he was cynically aware of just what would happen to their valuable reports when they did return to Russia.

Turning about, he set sail back along the coast towards Kamchatka. By the end of August his men were starting to die from scurvy. Water was running very low and strong head winds made the going slow They passed along the Aleutian Islands, and then came again to open sea. When land next appeared on November 4 it was thought at first to be the Kamchatka mainland. Storms had already severely damaged the ship and it was decided to land there rather than continue on to the home port of Petropavlovsk. But no sooner had the surviving members of the crew come ashore than a gale broke the ship's moorings and she was wrecked on the rocks.

It was soon found that, instead of the mainland, they were on an island—the island which has since been given Bering's name. Bering died early in December and so did thirty more members of his crew before the rest found they could subsist on the island's abundant wildlife, including seals and sea otters. They had found the home of the animals which had so intrigued the hunters in Siberia. The following spring, the survivors managed to build a small boat from the wreckage of the *St Peter* and returned to Kamchatka, where they had been given up as lost. Steller, who had at one point saved the lives of the entire expedition by giving them an infusion made from various antiscorbutic plants he had gathered, brought with him a treasury of information about the animal and plant life of the newly discovered land, which was named Russian America. The crew brought with them the furs they had gathered during their long winter on Bering Island—the blue fox, the fur seal and the sea otter amongst them. The 900 sea-otter pelts alone in trade with the Chinese were found to be worth a quarter of the cost of the entire expedition from when it left St Petersburg four years previously. By now the good-natured but unlettered Tsarina Elizabeth was on the throne, a daughter of Peter's but with none of his interest in science. All the scientific reports, logbooks and charts of the expedition were filed away in bureaucratic pigeon-holes and forgotten. Elizabeth had no interest in the land which Russia could now claim. The furs, on the other hand, caused wild excitement amongst the traders and merchants, especially when they heard of the large number of animals that existed. With no government to establish any kind of order, they had Russian America for themselves, to exploit by private initiative as they saw fit. Steller's remarkable observations, which gave the world its first account ever of the fur seal, the sea otter, the sea lion and the sea cow, as well as the spectacled cormorant which has since become extinct, were not published until 1751, five years after his death in Siberia. He never returned to Europe and spent the last years of his life continuing his investigations in Kamchatka. It was years before his reputation was rescued from oblivion, for the only concern now was the hunt for furs in the new world.

The next forty years saw a reign of terror spread from Bering Island along the Aleutians. Singly at first, then in whole fleets, came the boats of the Russian fur hunters—the *promyshleniki*, as they were called. They sought only wealth—and they found it. One voyage could make its promoter rich for life. Any of the Aleuts who resisted were killed out of hand. There was no attempt to make a permanent settlement. The hunters would arrive, slaughter as many animals as they could find, then return to live fatly on the proceeds. When promoters could not find professional crews for their boats they press-ganged serfs and peasants and bribed officials to release prisoners to them from the Siberian jails. Such men were not likely to show much concern for the natives. It was not long before they hit on the happy idea of enslaving male Aleuts to do their hunting for them while they lived like sultans in the villages with the Aleut women. In return they brought taxes. The only decree that Elizabeth ever thought to impose on Russian America was in 1748 when she decided the Aleuts should pay tribute to the government and sent Cossack collectors to gather it in. The Aleuts had never heard of such a thing as tribute, and it seemed to provide little protection in return. One collector, amongst other crimes, was known to have caused the deaths of over twenty Aleut girls.

In the early 1760s the natives of Unalaska and the neighbouring islands rose in revolt, killing the crews of five ships. It was the first and last time they ever did so. The Russians took a terrible revenge, torturing and killing many hundreds of Aleuts and systematically destroying dozens of villages. The Aleut population never recovered from these reprisals.

When the Russians began their invasion the Aleutian coast was teeming with sea otters; fur seals were found by the million in their breeding grounds to the north near the islands named the Pribilofs; and great herds of sea cows were present at all seasons of the year. Within only twenty-six years one of these species, the sea cow, was extinct. The only naturalist ever to see one was Steller and it is only through him that zoology has an account of its appearance and habits. It was not long before the sea otter and the fur seal were in danger of going the same way. In the Pribilof Islands alone the herds of fur seal were cut down by 90 per cent within twenty years. During the breeding season a man could walk across the beach and club and slaughter as many baby seals as he wished, for they had no means of escape. Rotting carcasses and even pelts lay everywhere, for much of the killing was done merely for the seals' sex glands, which in dried form were believed by the Chinese to have rejuvenating powers. This wanton slaughter was alleviated only by a falling off in the trade with China, due to political disputes, and the fact that the European markets became glutted with the furs that the Chinese refused to buy. By then, however, American and British fur hunters had come on to the scene and the slaughter continued until well into the twentieth century.

For the first twenty years of the Russian advance into North America no one in the West had any idea of what was happening. It was assumed that nothing had taken place after Bering's expedition after it had proved the main question that the two continents were separated by water. Foreigners who came to Russia were told that the sea-otter furs came from Siberia. But in 1762 Catherine II (the Great) came to the throne after a palace revolution and although she was the first competent ruler of Russia since Peter the Great and shared his interest in science and learning, she had her weaknesses. Lack of consistency was one—she changed her ideas and principles almost as often as she changed lovers—and boasting about the immensity of her kingdom was another. She couldn't keep the secret about Russian America for very long. The astonishment of the foreign ambassadors when Catherine told them may have been highly gratifying to her, but it provoked an immediate reaction from those countries with interests in America. Spain quickly occupied more of California and sent ships to explore further northwards. France sent an expedition under La Pérouse to see what the Russians were up to. And the British Admiralty sent James Cook with instructions to claim any uninhabited land on behalf of England.

England had long been searching for a North-west Passage between the Atlantic and the Pacific in order to be able to compete with Spain for trade in the Indies. The quest had started with Sir Martin Frobisher in 1576 and was continued by a long line of courageous explorers until well into the nineteenth century, including such men as Davis, Hudson, Baffin, Vancouver and Franklin. Some died in making the attempt and it was not until 1906 that Amundsen became the first to navigate the Passage in a voyage lasting three years. And it was not until 1969 that the tanker *Manhattan* became the first commercial ship to do so.

Cook tried to sail the North-west Passage from the Pacific in 1778, but was stopped by the build-up of ice in the Arctic Ocean. So he turned and set about charting the western and southern coasts of Alaska—and with such accuracy that few amendments had to be made to his charts by later navigators. They came to be used by the Russians themselves, including the English names given to many places, such as Cook Inlet and Prince William Sound. Cook made an attempt to claim some of the land, but it was really too late. By then the Russian fur traders, having depleted the seal herds along the Aleutian chain, were pushing further eastwards to find new sources of supply and this meant establishing a base of operations on the southern coast. At the same time the reported ill-treatment of the Aleuts was causing some concern in St Petersburg, where Catherine in one of her whims had decided that these new subjects should be converted to Christianity.

In 1784 an aggressively ambitious fur trader named Gregory Shelikov established the first Russian settlement on Kodiak Island at the mouth of the Cook Inlet. He was given a monopoly of the fur trade on the understanding that he would start a colony there and allow priests of the Eastern Church to come and teach the natives. This led in 1799 to the setting up of a new stockholder-owned corporation, the Russian America Company, which for the next sixty-four years, operating under royal charter, was alone responsible for administering and governing the new land. It was in many ways similar to other commercial companies of that period, like the Dutch East India Company and the Hudson's Bay Company, which were convenient devices for imperialistic expansion and exploitation. And it was from one of these, the Hudson's Bay Company, which had been operating in Canada since 1670, that Russia eventually faced the biggest threat to her American territory.

For the first nineteen years of its existence the Russian America company was managed by a man who became a legend throughout the Pacific—Alexander Baranov. He was without title or family background at a time when such things were all-important in Russia, having been born the son of a storekeeper in 1747, yet until shortly before his death in 1819 he ruled as absolute governor over the territory. It is unlikely that the Russians could have maintained their precarious foothold there without his tremendous energy and enthusiasm. From the initial base at Kodiak he established a series of trading posts and frontier stations and eventually built a new headquarters on one of the islands on the south-eastern coast, known first as Archangel. But the Tlingit Indians of that region had none of the mildness of the Aleuts and fiercely resisted the arrival of the Russians. The first stockade built there in 1800 was overcome in 1802 and its occupants massacred. Some years passed before Baranov subdued the warlike Tlingits. A second stockade was built and named New Archangel, gradually expanding to become the capital of Russian America.

The fortunes of the company rose and fell according to who was on the Russian throne. There were times, under Tsar Alexander, for instance, when the traditional Russian suspicion of foreigners was swept aside, the secret police put down, Russians allowed to travel abroad, and St Petersburg blossomed as one of the most cultured capitals of Europe. This was reflected in New Archangel, which became, as one writer described it, the Paris of the Pacific at a time when San Francisco was merely a Spanish frontier post. Street-walks were built, onion-domed churches, the houses furnished with elegance, and the evenings at Baranov's Castle, overlooking the town and harbour, given over to concerts and balls, a-glitter under crystal chandeliers with silken gowns and splendid uniforms. Ships from all over the world came to trade in what was the gayest and most sophisticated city of the Pacific. The streets thronged with American and British and French crews. There was everywhere the heady excitement of creating a new and growing empire. Scientific observatories were established to study the geology and mineralogy of the country and to chart the coasts. The native Indians were treated with sympathy and understanding. Not only were most of them converted to Christianity, but under suce men as Father Veniaminov and Governor Ferdinand von Wrangell hospitals and schools were established for them and many were able to take up senior posts with the company.

But, as her history shows, Russia was never quite decided whether she should look to east or west. There were other times, as when Nicholas I came to the throne, when Russia relapsed into gloomy, suspicious introspection. Foreigners were excluded—and this meant from New Archangel as well, whose harbour became empty and whose streets were deserted. The secret police once again held sway. In a close parallel to the present day few Russians were allowed to travel abroad because of the fear of outside ideas and influences and even movement within Russia was severely restricted. The nobles had always been against allowing serfs to go away from their villages and it was this attitude that was ultimately responsible for Russia losing hold of her American possession. Russian America was never really colonised. Even in the days of expansion the longest period of time that any Russian was permitted to stay there was fourteen years. He then had to return to Russia and his place taken by someone else. There were never more than a thousand Russians in the colony at any one time, and however much they might have wanted to, they could never make it their permanent home.

The hunters and traders of the Hudson's Bay Company, Scotsmen many of them, were under no such restrictions and they had gradually been colonising eastwards across Canada. In 1825 a convention was signed between Britain and Russia defining their boundaries in north-western America and establishing the important southern border at latitude 54° 40′. But in the absence of Russian posts in the interior and along the Yukon the men of the Hudson's Bay Company made frequent forays into that area, bartering with the Indians for furs. Since Russia at that time was beset by political troubles at home, and in any case was going through a period of withdrawal which resulted in a relative lack of interest in her American lands, little was done to prevent this. In fact, the Russians even agreed to lease the south-eastern mainland to the company for a ridiculously low rental of 2,000 land-otter skins a year. This lease lasted for twenty-eight years until 1867. By then, Russia and Britain had been at war in the Crimea and the Russians feared that if hostilities were renewed they would not be able to prevent the British from taking over the entire territory. The Russian America Company had not been doing well, and a sale of the land would enable a satisfactory liquidation of the company. If that sale could be made to the United States, with which Russia had established cordial relations, it would ensure that the territory was occupied by a friendly power. And it would forestall another possibility which some Russians were uneasy about: that the United States might decide to take it over regardless. It was an understandable qualm in view of the preaching of Manifest Destiny that was then widespread in the United States, urged as a holy cause by some political leaders.

And so, after long and secret negotiations between the US Secretary of State, William H. Seward, and the Russian Minister to the United States, Baron Edouard Stoeckel, the Treaty of Cession of Russian America to the United States was signed on March 30, 1867. The price reluctantly agreed by President Andrew Johnson and his Cabinet was $7,200,000. Russian America became Alaska and embarked on a new path as part of the United States, after 126 years of Russian occupation whose main effect had been to prevent it being acquired by Britain. It seemed to be the dawn of a great new future for this vast land, as later that year, on the afternoon of October 18, against the backdrop of New Archangel harbour, soon to be renamed Sitka, and the forests and mountains that ringed it, the double-headed eagle of the Imperial Russian flag was lowered and the Stars and Stripes of the United States took its place. But this was not to be. If the Russian occupation had begun with barbarism and never been more than tenuous, it had at least brought some semblance of justice and health and education to the Indians and for a while at least the flowering of a cultured civilisation at New Archangel. American rule, the replacement of a despotic government by one professing democracy, brought neglect at best and totally callous exploitation at worst. For the natives it brought new miseries in the form of liquor and syphilis. For those Americans who came to make their home in the new land it brought years of utter indifference on the part of succeeding Congresses in Washington until the long struggle to achieve statehood was at last won in 1959. The benefits went to outside commercial interests who made fortunes from furs and fisheries to the point where fur seals and sea otters became very nearly extinct and rivers ran empty of salmon. The cost was the death of a culture.

4
Arrival of Democracy

'Seward's Folly.' 'Walrussia.' '. . . an ice-box unfit for civilised men.' Such were the scornful epithets hurled by politicians and press alike at the news of Alaska's purchase after some adroit manœuvring by Seward to get the treaty ratified by the Senate. The only reason it passed at all was due to the friendly feeling that had existed towards Russia since her sympathetic attitude to the North in the Civil War. Many considered that the way it was done, in a hastily summoned executive session just hours before the adjournment of Congress, was an outrage of the Constitution, although the reason for such secrecy was not primarily due to public opposition but because of the need to take the British by surprise and foil their designs for taking over the territory. However, in using these moves as an excuse for attacking President Johnson's administration, the opposition condemned Alaska as a worthless and uninhabitable waste. The few who knew the territory hailed its purchase as a brilliant achievement. But they were in the minority. Most people knew nothing at all about Alaska and they weren't going to trouble to find out. From their hostility and ignorance was created the myth of Alaska as a useless desert of ice—a myth which was responsible for much of the neglect that followed.

It took Congress fifteen months of bitter argument to get around to paying Russia and there were times when it seemed the United States might not honour the deal. When the appropriation bill did eventually pass, it was only after Stoeckel, whose reputation depended upon it, had resorted to bribing a number of high-placed officials who were exposed after a Congressional investigation. President Johnson himself only narrowly escaped impeachment by Congress. And there was criticism in Russia as well. Angered officials of the Russian-American Company stirred up a storm of controversy in St Petersburg, accusing the government of selling peoples who had been converted to Orthodoxy. Suspicion soured the friendly relations that had previously existed between the two countries and they were never the same again. Both Seward and Stoeckel became discredited men as for one reason or another each country felt it had somehow been swindled by the other.

But in the jubilant ceremonies at Sitka to mark the handing over of Alaska, there was no hint of this trouble and no suspicion of what was to come. The city was a thriving community of nearly three thousand people with schools, scientific institutes, hospitals, a public library, churches and a theatre. Smallpox had been wiped out by widespread inoculations. The sea-otter herds were being restored by the conservation policies adopted by Wrangell. The future looked highly promising, especially when the Russians and those of the natives who had been educated were given the option of United States citizenship, to be protected in the free enjoyment of their liberty, property and religion. The thousand or so members of the now defunct Russian America Company who gathered in Sitka from all over the colony had three years to decide if they wanted to stay. At any time within that period they could choose to return to Russia with transportation provided. Few imagined that they would want to do so. Together with the American frontiersmen who flocked to Alaska to take their chance in this new dominion, they began staking out ground for homesteading. New stores and restaurants and saloons were built, a newspaper started, and the first steps taken to organise a city government. It was assumed that settlement would follow the pattern already established in the western territories and California. They were exciting, promising days. They were not to last for long.

Congress had been asked in approving the treaty of cession to attend to proper legislation for the occupation and government of the territory, without which no settler could acquire title to land. Indeed, only a week after the main handing-over ceremony at Sitka the inhabitants were reminded that attempts to acquire land were illegal and that they were in danger of being forcibly removed. With its ears still ringing from the caustic indictments of Alaska as a barren wasteland, Congress wanted as little to do with it as possible. Alaska was created a Customs district, which merely meant that the same duties on foreign goods applied as in the rest of the United States. The Pribilof Islands were made into a reservation and an exclusive twenty-year concession for their seal fisheries granted to a private company in San Francisco in return for a federal royalty on each of the 100,000 skins that were allowed to be taken each year. And those were the only two pieces of legislation relating to Alaska to be enacted for seventeen years. Incredible as it may seem, during the whole of that time of 'no government' Alaska was without laws. No one was legally entitled to be married or buried, to buy or sell, or to settle the land. Criminals could do as they wished with impunity, only having to fear reprisals from their victims or from the families of victims. The only legal violations were those under the Customs Act, which included prohibition of the importation and sale of distilled liquors and the sale of liquor to Indians. In such an event, violators were subject to prosecution in the neighbouring territories of Oregon or California. This was hardly likely when the only legal forces available were the military and they were the most lawless of all.

Since no government department in Washington wanted anything to do with Alaska, it was decided that the 250 American troops who had come to Sitka to take part in the handing-over ceremony should remain and that Alaska should become a military district, with posts at Sitka and Wrangell. This was intended merely as a transitional step until a proper form of government could be established. It lasted for ten years and then only came to an end when the troops, by then grown in number to 500, were withdrawn to help put down an Indian rising in Idaho. Their commander, Major-General Jefferson Davis (not the Confederate leader), had come out West to fight the Indians and he wasn't about to change his ideas just because he had been sent further north. He didn't think much better of the Russian foreigners in town. Within days his troops were on a drunken spree, raping Indian and Russian women alike and looting buildings, even the cathedral. If it had been the Aleuts who suffered most from the first arrival of the Russians, now it was the turn of the Tlingits under the American military occupation. Syphilis and drunkenness spread quickly amongst them, the liquor being smuggled in illegally by the military and also made from imported molasses and sugar. Davis and his officers themselves held debauched parties with the Indians. Any suspected slight was answered by shooting up a Tlingit village or two.

Complaints to Washington went unanswered. Bitterly the Russians began leaving Sitka within months of the transfer. Not for them the rights and privileges of American citizenship. By the end of the year hardly any of them were left in Alaska and even the Americans who had arrived with such high hopes had had enough. Sitka gradually faded into a ghost town. It was not entirely the fault of the military. Congress had given Alaska no laws, sent no ships, provided no one to staff the hospitals and schools. When the troops eventually left in 1877 there were no more than twenty families left in Sitka, of which only five were Russian. Alone and unprotected, they were now left to face the wrath of the Tlingits which had been building up to boiling point during the ten years of military misrule.

Desperate calls for help continued to find deaf ears in Washington, even those from the collector of Customs duties and his five deputies who were the only government officials in Alaska at that time. Their work was virtually impossible because of the lack of any revenue ships to prevent smuggling from Canada and the fact that correspondence with the Treasury Department invariably went unanswered. They had, for instance, sought advice about the importation of molasses which was being used for the manufacture of 'hoochenoo', but were ignored. The only decision taken by the Treasury was to recommend that the Customs district be abolished because the receipts were not worth the cost of administration.

With the Tlingits threatening a bloody massacre, the few left at Sitka turned to the British for help. An appeal addressed to the captain of any British ship at Vancouver was promptly answered by Captain H. Holmes A'Court, commanding HMS *Osprey*, who set sail without awaiting orders and arrived at Sitka on March 1, 1879, where he found the inhabitants in a great state of anxiety and alarm. He promised to remain until relieved by an American vessel and thus averted an almost certain Indian uprising. Shamed into action, the US federal government sent warships of its own, and they were to remain in Alaskan waters to give protection for the next twenty years. For the first five years, in fact, the commanders of US Navy vessels successively stationed at Sitka were the rulers of Sitka.

Elsewhere in Alaska things were more peaceful, notably in the region of the Pribilof Islands where the newly formed Alaska Commercial Company had purchased the assets of the Russian America Company and had obtained the sole franchise to hunt for fur seals and sea otters—curiously, after putting in the lowest of thirteen bids. The company was after profits, and big profits at that. But they saw to it that the Aleuts received at least some payment and were housed and schooled. They organised the only mail and passenger service in the territory. It was no panacea, but it was better than the total lack of administration in the rest of the territory. With the company paying some $300,000 a year to the federal government in royalties on the furs taken, it would be only twenty-four years before the seal fisheries alone had repaid the purchase price of Alaska. That was as far as the federal government, the absentee landlord in Washington, thought about it. The easy answer seemed to lie in granting to commercial organisations the right to exploit Alaska's natural resources and let them manage their own affairs as long as they returned a share of the profits to the national money-box. And so began the age of the monopolies which so bedevilled Alaska for the next ninety years. For if those in Washington saw little but a barren wasteland in Alaska, not so the businessmen and industrialists with a sharp eye to the main chance. Here was a land rich for the plundering. They were quick to see the opportunities that existed in furs, fisheries, canneries, lumber and, later, gold, copper and other minerals. If they were to have the free hand they wanted, and no local taxes to pay, they had to ensure two things: that Alaska was kept without a population and without a government. The best way to do this, sheer bribery notwithstanding, was to foster and maintain the image of Alaska in the public mind as an uninhabitable, Arctic desert. Congress had given them a good start anyway—it only remained to employ lobbyists in Washington to make sure the message wasn't forgotten. They worked to such effect on imaginations that were so lacking that echoes of that image are found even today.

'All this fuss—who cares anyway? There's nothing up there but snow and ice.' The oilman who says that in 1972 is only repeating what the fur hunters were saying a century earlier.

The Alaska Commercial Company was the first-comer. In the twenty years until its lease expired in 1889 and another Californian group took over it averaged over 90,000 sealskins a year and netted eighteen million dollars profit for its sixteen stockholders. By the end of that time the fur seals which the Russians had nearly exterminated but carefully built up again by conservation into a herd of some three million had become seriously depleted. These fur seal which have their breeding grounds on the Pribilof Islands, where they migrate every summer from the west coast of the United States, are distinct from other kinds of seal distributed throughout the rest of the world and whose pelt is relatively valueless. Their depletion was not only due to the company but to the highly wasteful killing at sea which was practised by ships of other countries, notably British and Canadian. Seals would be shot when coming up to the surface to breathe, but most of them sank before they could be gaffed. Since it was not possible to distinguish between male and female and most of the females were with pup, their death doubled the loss. Not that the intrepid hunters cared—again it was only profits that counted. The slaughter continued until the fur seal was close to extinction and it was not until 1911 that pelagic sealing (killing at sea) was abolished by international treaty and the governments of the

United States and Russia, the two countries with seal herds, took over the management of land sealing, allowing Great Britain and Japan 15 per cent each a year of the American product. The sea otter fared no better and was also close to extinction before it came under rigid protection in 1911, since when no hunting of it has been allowed.

It was not only the animals who suffered depletion. As the profits of the commercial companies diminished, so did their care of the Aleuts and the Eskimos with whom they had come in contact. The schools which the lobbyists had spoken of in such glowing terms were found by visiting Treasury agents to be virtually non-existent. Villages were insanitary and disease-ridden, the native population steadily declining as deaths exceeded births. But, in spite of recommendations that the federal government should make itself responsible for its new citizens, nothing was done. The leasing of the seal islands to private commercial concerns continued. The only real governmental interest was in whether or not the fur-seal herds of the Pribilofs should be protected. Arguments flared up in Congress on the subject, there were charges of corruption and perjury. With the furs bringing in millions of dollars in profits and royalties, paying many times over for the original purchase of Alaska, this was money, something the politicians could really be concerned about. The estimated three million seals that the Russians had left were reduced to less than a hundred thousand before action was at last taken to save the species.

Meanwhile, the Eskimos of the northern coastal regions, who had until now escaped the plight suffered by the Aleuts and Indians from Russians and Americans alike, were facing a new and grave peril. Ever since inhabiting these lands, indeed the only way that living there was made possible, they had relied on whales and walrus to provide food for themselves and their sled-dogs and skins for their boats and articles of clothing. Nothing of these animals was wasted. On the other hand they were also sought by Yankee whalers from the ports of New England—the whale for its oil, used as lamp fuel before the advent of petroleum and for its baleen from which 'whalebone' corset stays were made, and the walrus for its ivory tusk. The carcasses which were of such value to the Eskimos were thrown overboard as waste. With the coming of steamships in the 1880s the killing became more intense and extended further northwards. The walrus herds, like the seals, neared extinction, whales became scarce as they were driven from the coastal areas. The Eskimos faced starvation—and the federal government was not going to lift a finger to help them. They were too far away in that northern ice-box for anyone to be able to comprehend them or their problems. They were those quaint savages who rubbed noses and swapped wives and lived in ice-built igloos—so went the stories of the explorers of the time, garnishing their exploits with half-truths.

But there were a few who cared. Just as the Russians had converted many of the Aleuts to the Eastern Orthodox Church and had at the same time helped them materially, so selfless and well-intentioned American Protestant missionaries were doing similar work among the Eskimos, founding schools which were supported entirely by private donations. In recent times, missionaries have been criticised by some for brainwashing and confusing native peoples and changing their way of life which might have been primitive but which had its own justice and standards—more so indeed than many of those professing Christianity. This was certainly so in the case of the gentle Eskimos. But the fact is that without the endeavours of the missionaries the 15,000 Eskimos living in Alaska at that time would have died out, as did so many of the Indian tribes in the rest of the United States. Their plight aroused the sympathy of the Reverend Sheldon Jackson, general agent for education in Alaska, during his first visit along the shores of the Bering Sea and the Arctic Ocean in 1890. He had seen similar semi-nomadic peoples on the Siberian coast living well off herds of domesticated reindeer and felt that these animals could solve the Eskimos' problem. Instead of pauperising them with state charity, import reindeer and let the Eskimos support themselves.

Everyone, including the Eskimos, thought it was a good idea. Congress considered an appropriation of $15,000 to introduce domestic reindeer into Alaska and thus help the starving Eskimos. Even this paltry sum was too much. Congress adjourned without passing the measure. This shameless neglect spurred Jackson to seek private funds from churches and from the general public, writing to the press and speaking from the pulpit about the problem. Of the natives he wrote:

'Russia gave them government, schools and the Greek religion, but when the country passed from their possession they withdrew their rulers, priests and teachers while the United States did not send any others to take their place. Alaska today has neither courts, rulers, ministers, nor teachers. The only thing the United States has done for them has been to introduce whiskey.'

In such a way Jackson began to arouse the American conscience. With private contributions of $2,000 he imported sixteen reindeer to the Seward Peninsula as an experiment. It worked. More animals were brought in, again using private funds for it was three years before Congress contributed a miserly $6,000 to the scheme. They began to procreate until they were widely spread over the western region. The

Eskimos were saved, and for many years the reindeer were a mainstay of their economy.

After fur, the salmon fisheries of Alaska were the next to attract the attention of the fortune-seekers. Two canneries were established near Sitka and on Prince of Wales Island in the south-eastern region in 1878. Ten years later there were thirty-seven, spread out along the entire southern coast to Bristol Bay. It was too easy. For a few weeks each year, as the salmon made their runs upstream to the inland spawning grounds, the rivers were swollen with fish. All one had to do was to erect a barricade across the river and net as many fish as required, the only limit being the size of the available canning facilities. This meant of course that few fish would get through to spawn, but the immediate profits were too tempting for anyone to think of the future. The catch in 1889 of 700,000 cases was worth nearly three million dollars. But even before the industry had become fully established the seeds of its future decline were sown by reckless overfishing. Various attempts were made to control fishing—that very year the erection of barricades in Alaskan rivers to prevent the ascent of salmon to their spawning grounds was declared unlawful. But Congress provided no means of enforcing the law and this was to be the story for the next sixty years. The fortunes of the industry rose and fell as rivers became depelted and were abandoned until after a number of years they gradually filled with fish again and then the whole cycle was repeated. One of the peak years was 1918, when 135 canneries packed over six and a half million cases worth more than fifty million dollars. By 1921 the catch was less than two million cases. With the mid-1930s came the biggest catches in Alaska's history, reaching nearly eight and a half million cases in one year and establishing Alaska as the world's principal salmon producer. Twenty years later the fisheries had declined to such a point that in 1953 Alaska was declared a 'disaster area' by President Eisenhower, a term normally used for such major calamities as hurricanes and earthquakes, and the territory received federal relief funds. Since then the catch has slowly been building up, but is still far lower than the record years.

The long period of federal mismanagement and congressional inaction saw tne by now familiar arguments between the need for conservation and the desire for quick profits. But even in the good years it was not Alaska or the people living there who benefited. The canneries were owned almost entirely by outsiders, mainly from California and the West Coast, and even the workers they used for the short season were brought in from outside, cheap Chinese labour being plentiful. In the early days no government in Alaska meant no local taxes to pay, so the fishing and cannery interests had their own reasons too for promoting the image of the worthlessness of Alaska. Meanwhile, just as with the Eskimos to the north, the Indians of the southern coast were deprived of the fishing that had always been one of their main sources of food.

In spite of efforts to discourage people from going to Alaska, the population was slowly beginning to grow as prospectors and other adventurers drifted northwards. A census taken in 1880 showed there to be 430 white people, less than half than under the Russians, and 33,000 natives of whom half were Eskimos. Those who wished to settle still could not homestead or even buy land from the government, and the miners themselves could not legally stake out claims. Alaska was an extraordinary kind of no-man's-land. No less than twenty-five bills providing for civil government had been introduced into Congress in the fifteen years following the purchase, but none had got even as far as being debated. The pressure on Washington was mounting, however, particularly with the publicity being given to Sheldon Jackson's reindeer scheme for the Eskimos, and at last, in 1884, the first Organic Act was passed. It provided for a governor and a district court and the appointment of various officials. But it was so ambiguous and badly drafted that it was little better than useless. The homesteading provisions included when it was first drawn up were deleted after pressure from the fur and fisheries lobby. Alaska was still a land without land laws. For no particular reason, the laws of Oregon were made to apply to Alaska. These referred to such establishments as counties and towns which simply did not exist in Alaska. Even more farcical was the fact that under the Oregon code only taxpayers could sit on juries, but since the Act did not provide for taxes to be raised in Alaska, there were no taxpayers and therefore no legal juries could be formed.

The confusion was complete. Murderers walked around scot-free. Prohibition was unenforceable because of the legal tangles and even more liquor found its way to the natives. No legislative body was created, and there was no provision for the election of representatives to go to Washington where they could at least have explained the situation. The $25,000 a year provided for education was pitiful; even the Russian government, through its church schools which were still operating in Alaska, was spending more. And yet, by this time, the furs and fisheries of Alaska had contributed more than thirty million dollars to the national wealth and more than repaid the Treasury for Alaska's purchase price. And soon there was to occur an event that would ensure Alaska the attention not only of the United States but of the world. An event which fired the imagination of men and brought thousands stampeding to the northland. The discovery of gold.

5
From Gold Rush to Statehood

The Klondike gold rush began on a warm summer morning at the end of July 1896, when the rivers were swollen with melted snow and men coated their exposed flesh with clay to keep the mosquitoes from biting. George Carmack's Indian squaw had cooked breakfast as usual and taken the skillet to wash in the nearby creek. Carmack, a Californian, who like many other prospectors at that time had married an Indian girl, sat outside his tent with her two brothers, Skookum Jim and Tagish Charlie. They were camped by Rabbit Creek, near the point where the Klondike River flows into the Yukon. Earlier that year they had been wandering along the Yukon on the Alaskan side of the border, prospecting and trading furs. Now they had crossed over into the Yukon Territory of Canada.

Sudden shouts from Kate, the Indian girl, roused them from their early-morning smoke. They ran down to the creek, and there, in the pan which she had been washing out with gravel and water, sparkled tiny nuggets of gold. They stared, scarcely able to believe it. Then without a word they rushed back to get their own pans and feverishly began sieving the dirt and gravel from the bed of the creek. More gold showed. There was no doubt about it. They had struck it rich. Just how rich was revealed a year later when a ship docked in Seattle with what the world's newspapers heralded as a 'ton of gold'. And so started the Klondike legend and the great gold rush.

The news was not long in spreading to other prospectors at the Sixty Mile and Forty Mile camps on the Yukon and to Eagle and Circle

City in Alaska. The creeks flowing into the Klondike were quickly staked out and a tent camp mushroomed into what was to become the new frontier city of Dawson. But the 'ton of gold' story that flashed round the world kindled the fever of many who never before had even considered prospecting. They came in their thousands, lured by the thought of gold just lying around to be picked up. The main routes were through Alaska, either up the Yukon from the Bering Sea or over the cross-country trails from Skagway. A few became rich—many more perished through weakness or lack of the necessary equipment. The northland has never taken kindly to those who came ill-prepared or who showed little respect for her changing moods. The Indians coined a name for such greenhorns—Cheechako—which was the closest they could get to the word Chicago. And others came too, in the extravagantly ornate stern-wheeled river boats chugging up the Yukon and Tanana rivers—preachers and dance-hall girls, whiskey drummers and shopkeepers, pimps and gamblers. Those who couldn't make it all the way to the Klondike spread out over the vast interior and western regions, digging potholes and panning every stream they could find. Within three years big strikes at Nome and in the Tanana Valley turned the Klondike into the Alaskan gold rush. New trails were broken, cabins built, villages and settlements sprang up. It was the biggest thing that had ever happened to Alaska, recorded by such writers as Jack London, Robert Service and Rex Beach as the last chapter in the great American saga of 'the West'. Millions of people in the United States and throughout the world came to hear of Alaska for the first time. It could no longer be kept as the private preserve of the fur traders and the fishing and canning industrialists.

There had in fact been gold strikes before in Alaska. The first and biggest was in 1880 when two miners, Joseph Juneau and Richard Harris, found gold in the upper reaches of the Inside Passage leading to Skagway in the south-east region. Thus was created Juneau City and the Treadwell mine, which in twenty years became the largest gold mill in the world. Circle City was founded in 1894 on the banks of the Yukon after gold had been discovered in the streams nearby, but it took the Klondike to catch the imagination of the world.

Alaska at that time, even after the Organic Act of 1884, was still a lawless land. Because of the complexities of the Oregon code the laws were difficult to apply and in the interior, anyway, there was no one to apply them. So the miners organised their own law. Each camp meted out justice by majority vote. Hanging was the punishment for murder, banishment or whipping for lesser crimes. It was simple, rough justice —but it was all there was. After the Klondike strike, however, Congress was forced to sit up and take notice. For one thing, the 60,000 people who had rushed off to look for gold in Alaska were mostly voters from other states—their opinions mattered. In 1898 Alaska was for the first time given its own criminal code and judges appointed to the various district courts. It was somewhat unfortunate that this coincided with the discovery of gold along the beaches of the Seward Peninsula near Nome by prospectors who were too late to go up the Yukon because the winter's freeze had set in.

The gold hunt was well under way by the time Judge Arthur H. Noyes was appointed to the newly created Second Judicial Division at Nome in mid-1900, its worth that year totalling four million dollars. over the whole of Alaska in fact, during the twenty years since the Juneau discovery, gold production had yielded nearly seventeen and a half million dollars. Noyes and his confederate, a political boss named Alexander McKenzie, hit on a fraudulent scheme to take over the best of the gold claims by passing out illegal orders and injunctions in court. Local lawyers were bribed with shares in the Alaska Gold Mining Company which McKenzie had formed and which quickly 'owned' most of the mines in the area. The nearest federal court to which the victimised miners could appeal was in San Francisco, 3,000 miles away. By the time they got together to do so, the Company had made a fortune from the easily mined claims. Confident in their success, however, Noyes and McKenzie overreached themselves and were eventually arrested in what became a *cause célèbre*. Judge Noyes got off with a mere fine and McKenzie's one-year prison sentence was quashed by the intervention of his friend President McKinley. It was not a promising start to the introduction of law in Alaska.

One further piece of legislation before the end of the nineteenth century at long last extended the homestead laws to Alaska. But here again it was done in a half-hearted way. The 160 acres that were allowed elsewhere in the United States for homesteading were reduced to eighty in Alaska. Also, the law was drafted as if to imply that Alaska was an agricultural area, where one had to farm and cultivate the land to be able to prove title to it. Alaska was immensely rich in many ways, but it was not, except in a few areas, an agricultural country. The homestead laws were a source of contention for more than fifty years afterwards and one of the reasons why Alaska was never more than sparsely populated. It was an example of legislators governing in ignorance and from too far away to understand the local problems. And yet no one could be in any doubt as to the territory's worth. A Congressional subcommittee in 1903 reported that since the cession Alaska had netted the federal Treasury a million dollars more than all its expenses, including the original purchase price, and added to the wealth of the nation fity-two million dollars in furs, fifty million in fisheries, and thirty-one million in gold.

Even the distances involved in travelling Alaska were too great for outsiders to comprehend. A journey from Sitka to the Yukon, for instance, began with a 166-mile voyage in the monthly mail steamer to Juneau. A canoe and natives then had to be hired to take the traveller and his provisions another 100 miles to the head of the Inside Passage. Here he would leave the water and hire more natives to trek on foot over the dangerous mountain trail to the upper reaches of the Yukon. He would have to arrive there between May and October because the river was frozen over for the rest of the year. To visit anywhere along the 2,000 miles of the Yukon to the Bering Sea a raft would have to be constructed on which the traveller could float downstream. By the time he reached a settlement like Anvik, he would have travelled for two months and over two and a half thousand miles. To reach Bethel, on the other hand, only 150 miles south of Anvik, he would have to take a steamer to San Francisco, wait for a ship that happened to be sailing to Unalaska in the Aleutians, from there take another vessel to the Kuskokwim River, and complete the four-and-a-half-thousand-mile journey in a seal-skin canoe. Such a journey might take six months or more. And yet under the rules laid down by the federal government the general agent for education was supposed to visit each school district in Alaska at least once a year. It was, of course, an impossibility. It might take a whole year to visit just one school in a very remote northern area.

Gold production in Alaska reached its peak in 1906, the main mining area then being centred at Fairbanks on the Tanana River. Some 6,500,000 ounces were mined that year, nearly a quarter of the nation's gold. Thereafter production declined as the more readily accessible areas were played out and the lack of roads or trails made it difficult to open up those interior regions that could not be reached by sea and river. In summer horses and men laboriously hauling equipment would sink in the marshy ground; winter travel by dog teams over frozen ground and rivers was faster and a good deal cheaper, but carried with it all the dangers of exposure or being caught in blizzards. However, more than gold had been discovered in the great drive northwards. Miners found silver, copper, platinum and coal in abundance, and with all the publicity that Alaska was now receiving it was a stock promoter's paradise. The great capitalists and bankers of the day also became interested. Dominating the entry of the big outside mining interests was the Alaska Syndicate, formed in 1906 by J. P. Morgan and the Guggenheim brothers to develop the giant Kennecott copper mines 190 miles inland from Cordova. The Syndicate built a railway from the Copper River Valley to bring out the ore, and also had major interests in steamship lines and canneries.

It was to some extent due to this new period of industrial growth that Congress took the first tentative steps towards providing Alaska with self-government. The introduction of a legal code in 1898 had also brought taxation but no provision for Alaska to be represented in Washington. There were fierce protests about taxation without representation, but even when in 1906 one delegate from Alaska was permitted to sit in the House of Representatives, he was not allowed to vote. It took another six years of struggle to win official territorial status by the passing of the Second Organic Act of 1912, granting Alaska its own legislature—a privilege given to Hawaii and Puerto Rico in 1900, only two years after they had come under the US flag. But every act passed by the legislature was subject to Congressional veto, a discrimination never applied to the other mainland territories which had by now all become states.

In 1915 a start was made on Alaska's first public railroad, from Seward on the Kenai Peninsula to Fairbanks, one of its specific purposes being to open up the important coalfields. The US Navy in the Pacific saw that it would urgently need this coal in the event of war, for the Panama Canal was not yet completed and fuel supplies were having to be shipped all the way round Cape Horn from the East Coast. Camps for the construction workers were set up at the head of Cook Inlet, and thus Anchorage came into being. There were schemes for more railroads, docks, telephone and telegraph lines. The boom days of the gold rush were petering out—days which had brought very great suffering and very few riches to most of the gold-seekers. Ten thousand left Alaska between 1910 and 1920, reducing the white population to about 28,000. But there were signs now that Alaska would be able to build more solidly on its natural resources. It was a time of expansion and opportunity, similar in a way to the best days under Russian rule. The progressive feeling in the air was represented by the very first official act of the new Alaskan Legislature, which in 1913 gave women the vote, seven years before it was generally adopted by the rest of the United States. A wide range of liberally-minded social and conservation measures were also passed from the capital, which had now been moved from Sitka to Juneau, and plans to help poverty and disease amongst the natives resulted in a gradual increase from then on in their population, at that time around 25,000. Where roads didn't exist, the bush pilots were blazing new trails across the sky and opening up even the most remote regions.

But most industries were still largely in the hands of outsiders. And with the granting of territorial status the federal bureaux had moved in, each jealously guarding its own particular interests whether it be forests and farming under the Department of Agriculture, furs and fisheries under the Department of Commerce, or wildlife under the Department of the Interior, with so many of these overlapping that Franklin Lane, the enlightened Secretary of the Interior in Woodrow Wilson's cabinet, published a critique called *Red Tape in the Government of Alaska*. The problems arising from this bureaucracy were irritating and did little to help efficient development, but a steady growth was achieved in the first two decades of the century. Unfortunately, the two factors of outside exploitation and federal management coincided in the early 1920s with an era of political scandal in Washington under the corrupt Harding administration, culminating in the Teapot Dome affair in which large areas of public land were virtually given away to powerful oil interests. The people of the United States and, belatedly, Congress were suddenly made aware of what the 'robber baron' industrialists had been up to during the great drive westwards across America. There were scandals about the railroads, about the vast forest lands that had been destroyed in Oregon. Even President Harding, shortly before his death and after a visit to Alaska, said in a speech in Seattle that the 'looting of Alaska' should end. The shock and indignation of the public resulted in a country-wide demand for conservation.

In many cases it was too late to do much about it. The damage had been done. But development in Alaska was only just getting under way—and the federal bureaux, sensitive to public opinion, were in a position to enforce that opinion. Mineral leases were cancelled, and large areas of Alaska established as national forests. By the time the government-owned railway from Seward to Fairbanks was completed in 1923, Alaska was facing a slump which was to last throughout the 1920s. While other areas of the United States forged ahead expansively, Alaska remained in a cocoon of conservation laws, for in the public mind it had come to epitomise the wilderness which no longer existed elsewhere. This special feeling about Alaska has never been lost. There is a marked similarity between attitudes then and the call for conservation now which is so hampering the efforts of the oil industry in Alaska.

Conservation in the United States was certainly needed. It was a tragedy that it had come so late and after so much irreparable damage had been done. And a tragedy too that in general it came to be forgotten until the new awareness of it now, nearly fifty years later, when the harm caused has been so much greater. Although it was not the purpose at the time, the early neglect of Alaska had the effect of ensuring that it did not suffer the fate of other territories which were developed too quickly and with too little thought for conservation. Alaska is a wilderness still for a curious variety of reasons, not least of which is the fact that for such a long period it was so distant from the seat of government.

But it is easier to insist that others practise conservation than to do so in one's own environment and to one's own discomfort. It was all very well for the American public after the scandals of the early 1920s to pigeon-hole Alaska as a wilderness, but as events since have shown, they did little to prevent the creeping pollution of their own societies. And it was small consolation for the jobless and impoverished Alaskan to know that it was in the noble cause of conservation that he was kept in that situation. By 1930, Alaska's population had only fractionally grown to 59,000, well below what it had been in the first decade, while the total population of the United States had almost doubled since 1900 to 123 million. Alaska had become a convenient scapegoat for a national sense of guilt, and it only needed the oil industry at the end of the 1960s to talk about large-scale development there to revive echoes of that feeling. A closer reading of Alaska's history might have warned the oil companies to play the game differently.

The depression of the 1930s swept aside thoughts about conservation, and in fact helped Alaska's flagging economy through President Roosevelt's New Deal programme. His decision in 1933 to increase the price of gold from $20·67 to $35 an ounce revived the gold industry and brought giant caterpillar tractors to scour the placer creeks where before miners had panned and dug by hand. For a few years gold output was the highest in Alaska's history, peaking to twenty-six million dollars' worth in 1940. The same applied to coal and other mineral resources. Federal aid was poured into public works projects, such as airports, schools, bridges, and even a tourist hotel at Mount McKinley Park. Agriculture benefited in a special and dramatic way with the Matanuska colonisation project, which was again to put Alaska on the front page. Two hundred families from the depressed farming areas of the Midwest were settled at government expense in the Matanuska Valley at the head of Cook Inlet and given a new start in life, the aim being both to help them and to demonstrate the agricultural possibilities of Alaska. After a shaky start the scheme was a great success and established Matanuska as a major farming area where, because during the summer growing season there are twenty hours of daylight, produce grows to an enormous size. A forty-pound cabbage is not unusual.

Alaskans still resented so much outside control of their affairs, however, and there was a growing demand for statehood. This received a dynamic impetus when in 1939 Dr Ernest Gruening, the Director of the Division of Territories and Island Possessions, was appointed Governor of Alaska. With energy and political acumen he began a long and bitter struggle against the absentee landlords to present Alaska's case before Congress. Then the Second World War intervened and once again Alaska was discovered.

Ever since 1935 the US military and Alaska's delegates to Congress had been arguing for the establishment of air bases in Alaska. And as far back as 1904 the Navy Department had seen that a naval base in the Aleutians would be of great strategic value in the 'possible complications that may arise between ourselves and the Orient'. It would not seem difficult to understand. The Aleutians are only a few hundred miles from the northern Japanese islands, the Russian territory of Siberia only fifty-six miles across the Bering Strait. Yet when the war in Europe broke out in 1939 there was not a single military base of any kind in Alaska from which to defend the territory. And although the following year Congress approved the establishment of the Fort Richardson air base at Anchorage, it was far from ready for action when the Japanese struck at Pearl Harbour on December 7, 1941. Six months later, without meeting any counter-defence, the Japanese invaded the Aleutians and occupied the islands of Attu and Kiska, the only foothold they ever achieved in the Western Hemisphere. Some of the Aleuts captured were taken back to Japan. It required a costly campaign, including an extraordinary naval battle off the Komandorski Islands in March 1943, to drive the invaders from the Aleutians, during which 2,500 American lives were lost. But it did prove the strategic importance of Alaska, especially when the Cold War with Russia followed the defeat of the Axis powers. The continued military presence in Alaska, from a peak of 300,000 during the Second World War to a present average of about 35,000, became a major economic factor, responsible for the long-sought construction of highways and coast-charting. With the early-warning radar network in the far north, Alaska is now the United States' first line of defence against any Soviet aggression. The prediction by General Billy Mitchell in 1935, shortly before his death, that Alaska was the most important strategic place in the world and that 'he who holds Alaska will hold the world' has become truer than most people thought at the time.

The fight for statehood was helped by the return to Alaska of many young men stationed there during the war who had been impressed by its possibilities. They resented being second-class colonial citizens. They resented the deficient land laws which made homesteading so difficult. And they resented, too, the outdated legal system under which, for instance, the mentally ill were tried as criminals and jailed, a barbaric throwback to the Middle Ages. The natives of Alaska, the Eskimos, Indians and Aleuts, had plenty of resentments themselves, but these are dealt with in a separate chapter.

Statehood bills had been introduced at various times since as early as 1915 when Alaska was permitted to have its own representative in Congress, but had never got off the ground, any more than the efforts to obtain full territorial government. A referendum in 1946 showed without any doubt that a majority of Alaskans wanted statehood, and that same year President Truman unexpectedly supported the claim. But the bills continued to be blocked, primarily by the big industrial interests, and even in 1947 one lobbyist was still saying that 'the talk of Alaska's vast resources is greatly exaggerated'. By then the population had doubled over twenty years to nearly 128,000. As time went by, the question of Alaskan statehood became a national issue. A major drawback was President Eisenhower's initial opposition on the grounds of US defence needs and the smallness of Alaska's population, an argument which had long been used by the anti-statehood lobby. But mounting public opinion and the support of such men as Interior Secretary Fred Seaton gradually had its effect, and in fact the drafts of the various statehood bills became more generous with each successive Congress. When in May and June of 1958 first the House of Representatives and then the Senate finally passed the Alaska Statehood Act it granted privileges which had not been given to other Western states. The final measure in making Alaska the forty-ninth state came on January 3, 1959, when President Eisenhower signed the statehood proclamation. The torchlight parades which took place in Alaska were reminiscent of the celebrations ninety-two years earlier to mark its purchase from Russia. Alaska was now, within the federal restrictions that apply to all states under the terms of the Constitution, allowed to govern itself. Alaskans, for the first time, were permitted to vote in national elections.

The population of Alaska at the time of statehood was some 226,000, of which about 43,000 were natives. Of the state's total land area of 375 million acres only a little over a million acres had been patented or entered by private individuals, a comment on the unworkability of the homestead laws. All the rest was under the jurisdiction of the federal government until such time as it would become privately owned, just as public lands throughout the whole of the United States had once comprised nearly two billion acres, three-quarters of the total land area, but had been reduced over the years to about 400 million acres outside Alaska. Of the total federal land in Alaska, twenty-one million acres had been set aside as forest land, seven million acres as national parks and monuments including Mount McKinley, eight million acres as wildlife refuges, four million acres as native reserves of which three-quarters were regarded as Indian tribal lands, two and a quarter-million acres as military reservations, twenty-three million acres as a Naval Petroleum Reserve, several million acres for miscellaneous purposes by the many federal agencies, and a further thirteen million acres which were earmarked for withdrawal by the Bureau of Land Management in Alaska and nine million acres for the Arctic Wildlife Range. In all, some 105 million acres were reserved or pending withdrawal by the federal bureaux and not available to private acquisition.

Under the terms of statehood Alaska was given the right over a twenty-five-year period to select as state lands up to 103·5 million acres from the remaining vacant and unreserved public domain. Other provisions included the state's right to 70 per cent of the proceeds from the Pribilof seal industry and 90 per cent of the royalties from mineral leases on the public domain. The question was, which lands to select, for most of Alaska had never been surveyed. And so many problems arose to do with native land claims that in 1966 the federal government imposed a land-freeze on Alaska, halting any further leasing or selection, by which time the state had been granted patent to only five million acres and applied for another thirteen million. Because in the cheerful carve-up of Alaska between state and federal government everyone had ignored or conveniently forgotten one thing—that under the treaty by which it was purchased from Russia, the United States government had guaranteed to the natives possession of their own lands. They were claiming 290 million acres, later reducing this to forty million. The sins of previous neglect were truly coming home to roost, for no Congress had ever bothered to sign a separate treaty with the natives to deprive them of their lands as had been done elsewhere in the United States. And in other ways too the Alaskans were finding self-government a formidable business, once the initial enthusiasm wore off. Continued federal aid was still necessary to save the state from bankruptcy, accounting for about a third of its $150 million yearly income. But about the first selection of land made by the state there was no doubt. They were those areas which had been leased for oil and gas exploration. For a new factor was looming on the horizon, to be of greater importance than anything that had yet happened to Alaska.

PART TWO

The Oil Men

6

The Last Chance

By two in the afternoon it was already dark. The winter-pale sun had shown itself for only a few hours that morning, just barely visible above the rim of the horizon. There had never been any warmth in it, but as it sank into the Arctic twilight so the temperature had dropped even lower. It was now 45 degrees below zero. Metal burned at the touch, even through gloves.

A strong wind cut across the icy plain, sculpturing fantastic shadows in the snow-dunes, frosting the beards of the men returning to camp. As they entered the squat prefabricated huts, light spilled on to the tractors and trucks lined up outside, tethered like horses to electric power points or hitching posts to keep the heater fans going, engines left running to prevent them freezing up. When the doors closed they were swallowed up in darkness again. The only landmark in the frozen world of the North Slope was the tall spire of the drilling-rig, a lacework of girders lit like a Christmas tree. The nearest habitation was the Eskimo village at Point Barrow, 150 miles to the north-west. Just two miles northward was the shore of the Arctic Ocean. Not that there was any visible distinction between land and sea. Both merged in a continuous carpet of snow and ice. Only the maps gave the name of this part of the coast. Few people had ever heard of it before. Prudhoe Bay.

It was strangely quiet on the rig. After weeks of incessant roar, as the drilling bit was screwed deeper into the ground, the big diesel motors were now just ticking over. A continuous line of pipe, in twenty-foot sections fitted arduously together, had been sunk 8,883 feet

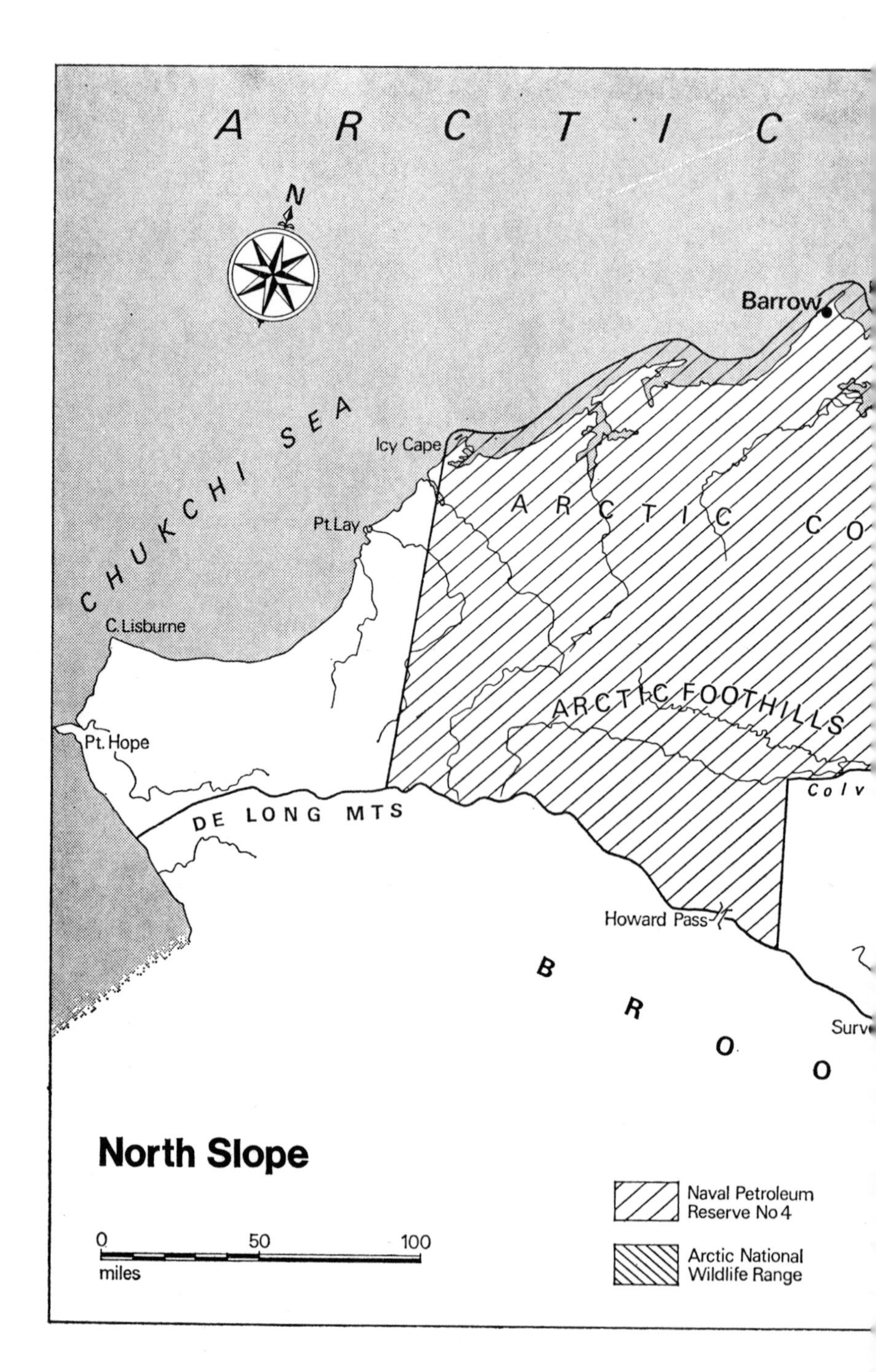
A R C T I C
N
CHUKCHI SEA
Barrow
Icy Cape
Pt. Lay
A R C T I C
C. Lisburne
ARCTIC FOOTHILLS
Pt. Hope
DE LONG MTS
Howard Pass
B R O O
North Slope
0
50
100
miles
Naval Petroleum Reserve No 4
Arctic National Wildlife Range

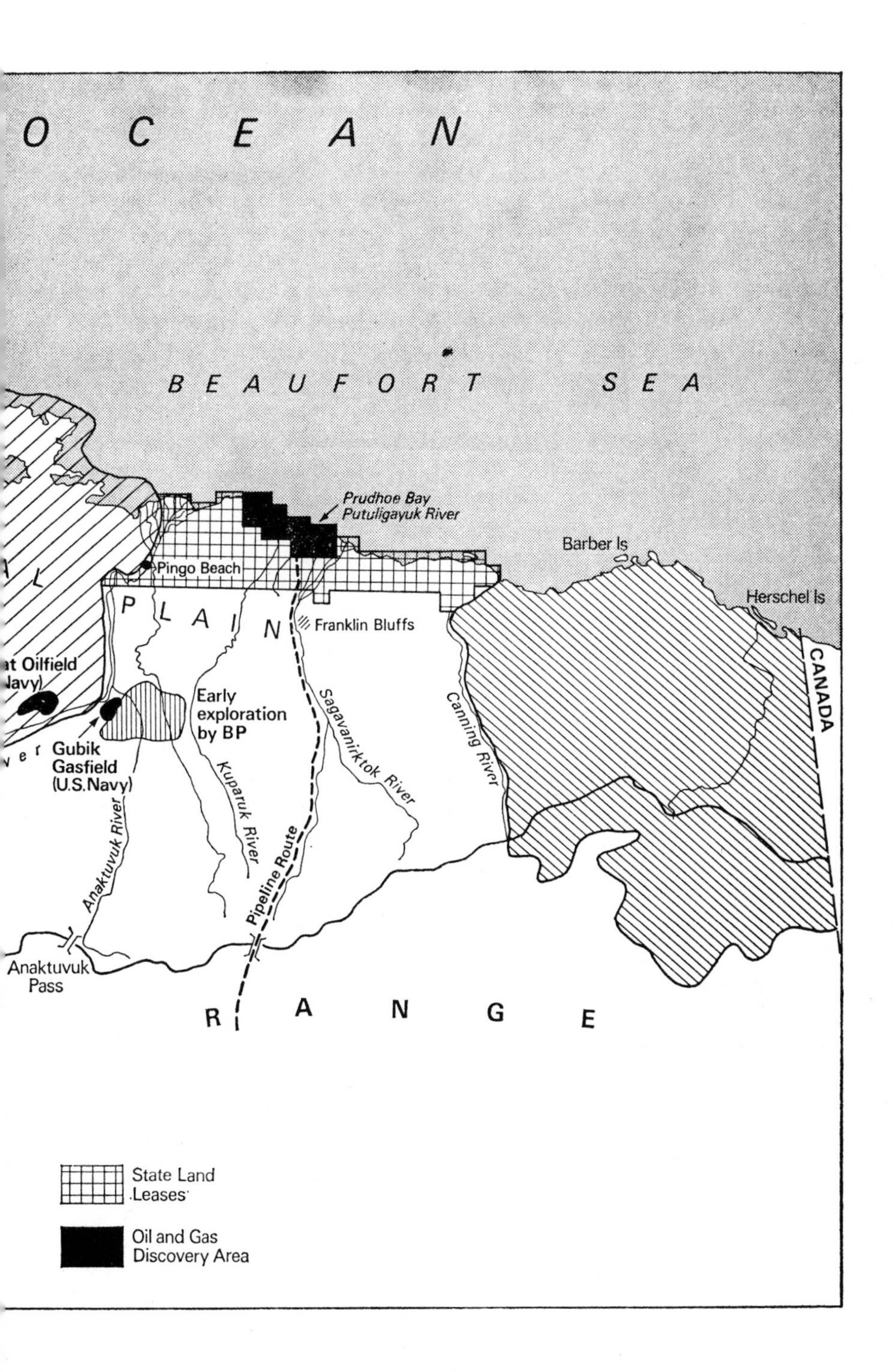
O C E A N
B E A U F O R T S E A
Prudhoe Bay
Putuligayuk River
Barber Is
Herschel Is
Pingo Beach
P L A I N
Franklin Bluffs
CANADA
Oilfield
Early exploration by BP
Gubik Gasfield (U.S.Navy)
Sagavanirktok River
Canning River
Kuparuk River
Anaktuvuk River
Pipeline Route
Anaktuvuk Pass
R A N G E
State Land Leases
Oil and Gas Discovery Area

down. There were only four men on the platform. Two were from Atlantic Richfield (Arco), the American oil company drilling the well: Jim Keasler, drilling engineer, and Bill Penteller, geologist. With them were William Stolland, the Loffland driller who had come on with the midday shift, and Ken Hughes, the test operator from Haliburton. They were about to test the only oil well operating on the North Slope. All the other companies who had been drilling there since 1963 had given up after failing to find any oil or gas. For Atlantic Richfield this well was a last-chance gamble. If it too failed, the company would leave as the others had done. It might be many years before they or others came back for another try. This well alone had already cost over four million dollars. It had been given the name Prudhoe Bay State No. 1. It was the last remaining symbol of the world's greatest and most powerful industry that was to wake Alaska from its long sleep. It was a symbol also of the hopes of the young state itself, for the land on which it was being drilled was amongst those few million acres selected by the state from what had been offered by the federal government. The date was February 18, 1968.

At 2.13 pm, by gently rotating the drill pipe, the valve of the testing apparatus which had been lowered to the bottom of the well was eased open. A rubber packer expanded against the wall of the hole, holding back the weight of the column of mud-like drilling fluid which normally circulated down the centre of the hollow drill pipe and up the outside to the surface, its purpose to wash out the fragments of drilled rock to cool the rotating bit, and to balance the pressure from any oil or gas that might be located. Thus the last 130 feet of the hole were opened up for testing. Any petroleum that might be present in the porous reservoir rock, trapped rather like water in a sponge, would now be free to flow up the drill pipe. All that controlled it in case the pressures were very great—and they could be thousands of pounds a square inch—was the weight of the drilling mud still in the pipe.

After seven minutes the returning mud began to bubble with gas. The valve was immediately closed off and preparations made for the main part of the test. There was natural gas present, they had known that much from two previous tests. The question was—how much.

The well had been spudded-in on April 22 the previous year, using a massive National 110 rig that had already drilled unsuccessfully the first well for a joint Atlantic Richfield and Humble Oil undertaking. Because of the enormous costs and uncertainties of exploration on the North Slope most of the early companies there had gone into partnership with others. Bringing in the dismantled National rig, which together with other drilling equipment, supplies and living quarters for the crews totalled more than 4,000 tons, had involved the biggest civilian airlift in Alaskan history. One Hercules sky freighter alone, flying in twenty-ton loads at a time, had made seventy-two round trips from Fairbanks.

Before that first well, Susie Unit No. 1, had begun drilling in March 1966, in one of the corporate shuffles that American industry was then undergoing a merger took place between Atlantic Refining and the Richfield Oil Corporation, a company which was already producing oil on the south coast of Alaska in Cook Inlet. By the time it was completed as a dry hole in January 1967, at a cost of four and a half million dollars, the new chain of managerial command was only just beginning to settle down. The new Atlantic Richfield Company still owned leases on the North Slope in partnership with Humble Oil, but the decision had to be made whether to abandon the search, as others had done, or continue in what was the most costly exploration the oil industry had ever undertaken in the most difficult and inhospitable of regions. The industry as a whole had already spent $125 million on the search by that time, with nothing to show for it. After some weeks of uncertainty it was agreed by the company's chairman, Robert Anderson, on the advice of exploration executives Harrison Jamison and Rollin Eckis, to drill just one more well. The Susie rig was hauled by caterpillar tractor to the site selected for Prudhoe Bay State No. 1.

'If that had been a dry hole I am certain that no future activity would have taken place for years to come.' So states Mr Jamison, the Atlantic Richfield geologist whose life has been monopolised by the North Slope. 'We had the only active rig drilling at Prudhoe and no seismic crews were active and no other drilling activity was contemplated. All prior results of drilling had been negative although some encouraging shows had been found. The main problem was finding a reserve large enough to justify economic recovery from this remote and harsh region.'

That was the crunch. It was not good enough just to find oil. There had to be enough of it to be worth producing commercially, for the cost of getting it out was going to be enormous. Of all the companies that had looked at the North Slope, British Petroleum was the most optimistic and had done most of the exploratory work, drilling eight wells since 1963, first in partnership with Sinclair Oil and then with Union Oil. These companies had both withdrawn when the search seemed to be unsuccessful. The British company had tried to go it alone, but because of a severe shortage of dollars it had to decide in mid-1967 to cut its losses and pull out. The costs were just too high in a region where every smallest component had to be shipped or air-freighted

therefore c
tive basis w
for a comp
not exceed
to a maxim
competitive
term, eligib
of fifty cents
most of these
leases are av
production is
ment—12½ pe
Alaska charge
This was origi
with current m
has provided sc
courage explora
5 per cent

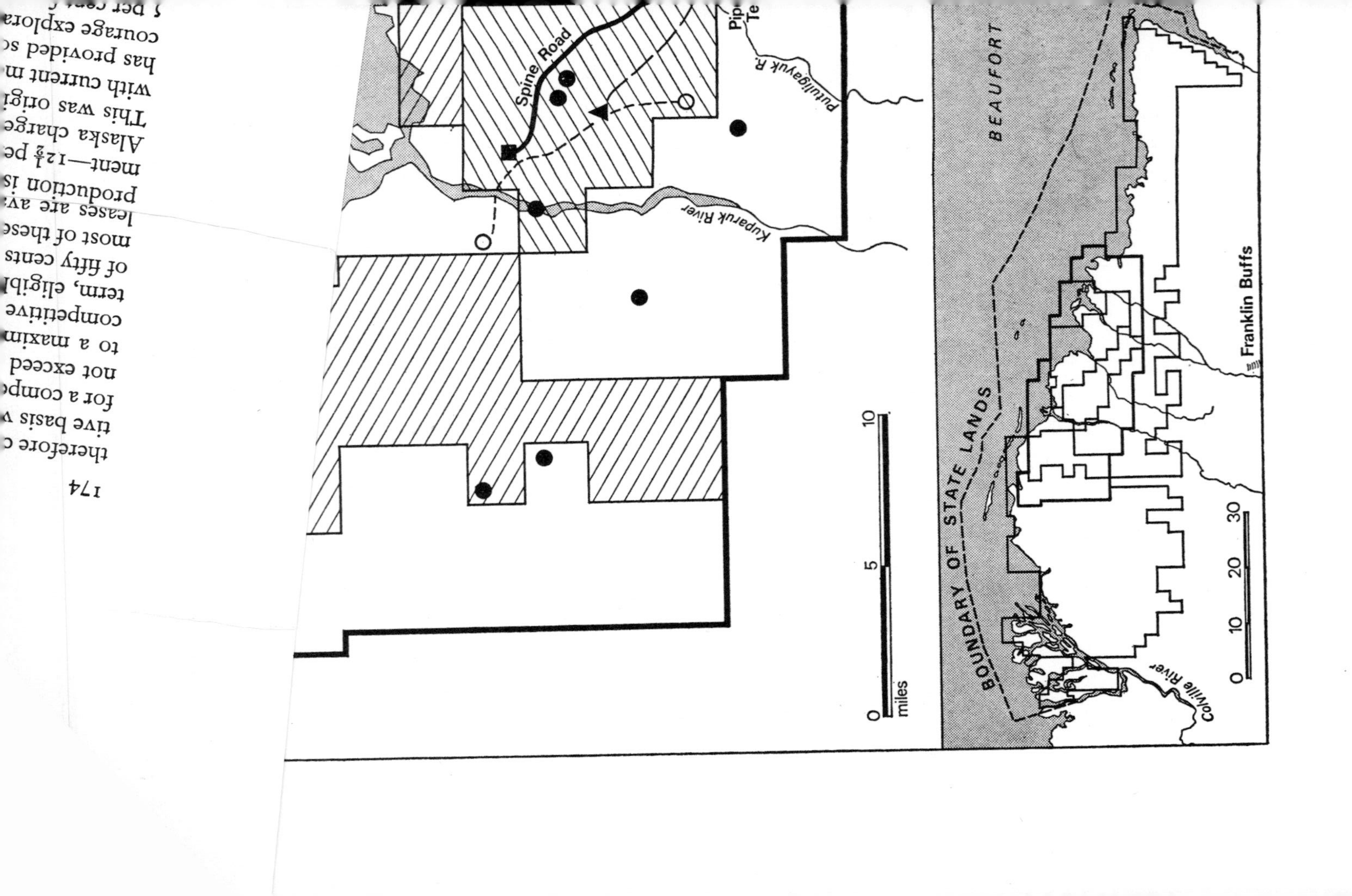

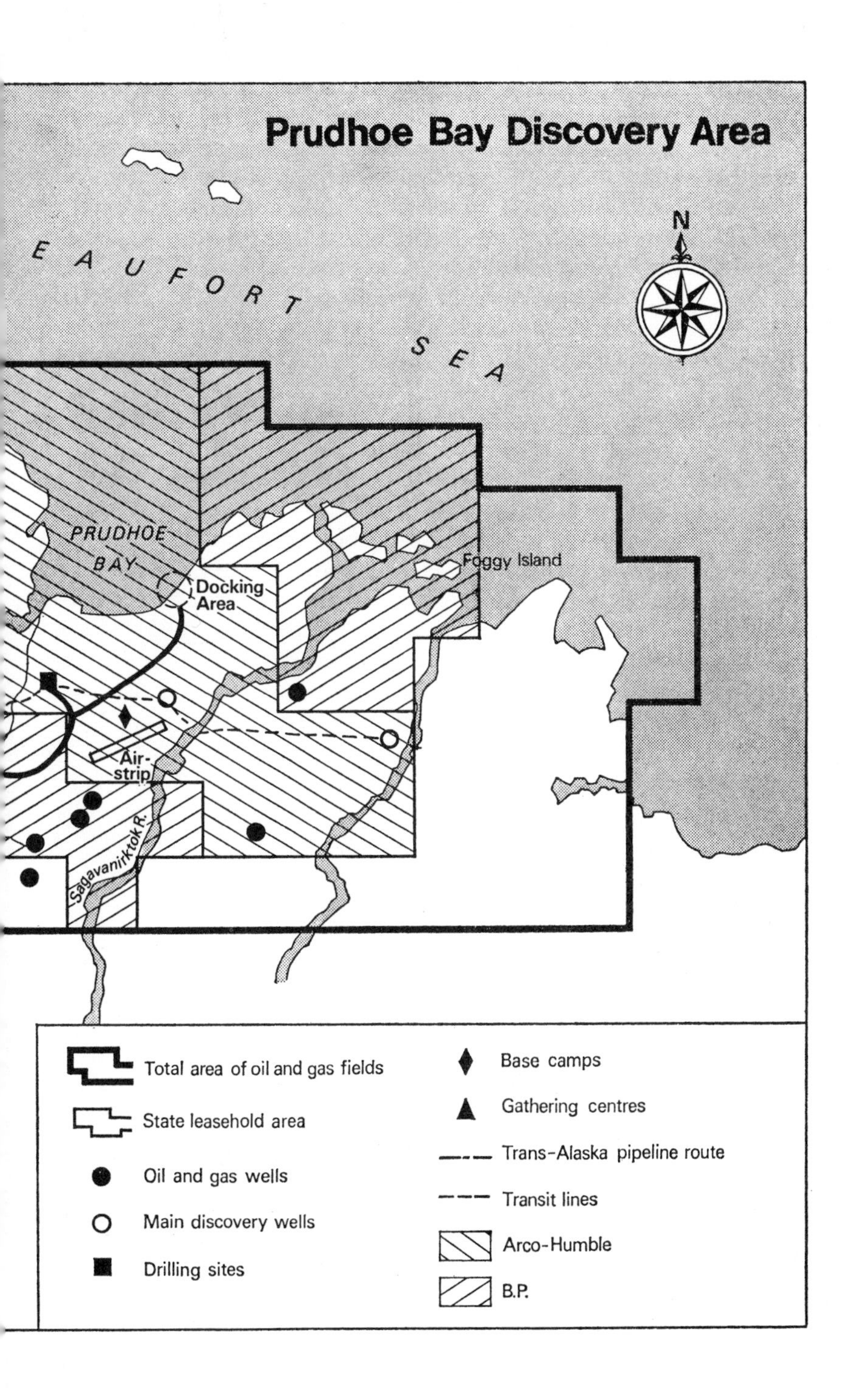
Prudhoe Bay Discovery Area
N
EAUFORT
SEA
PRUDHOE
BAY
Docking Area
Foggy Island
Air-strip
Sagavanirktok R.
Total area of oil and gas fields
State leasehold area
Oil and gas wells
Main discovery wells
Drilling sites
Base camps
Gathering centres
Trans-Alaska pipeline route
Transit lines
Arco-Humble
B.P.

from hundreds of miles away, where it was so cold that metal became brittle and snapped and men's efficiency was reduced by 80 per cent. However, the company still owned leases over large areas of the North Slope, bought for a minimal price when no one else was interested. These included thousands of acres at Prudhoe Bay and around the Colville River area. BP's geologists had selected these two places as being the most likely to contain oil. It was decided to drill at Colville first, since there the costs would be shared with Sinclair Oil. It could well have been Prudhoe—such is the element of luck in the oil business. But the company's Prudhoe Bay leases were to be of vital importance later.

The drilling engineer on both the Susie well and now Prudhoe Bay State No. 1 was Jim Keasler, a thirty-three-year-old Missouran who had come to Alaska in 1965 to work on the Cook Inlet rigs. By May 3 he and his crew had drilled down to 590 feet and run in the surface casing. Then, after an early break-up of the ice, operations were suspended for the summer. This was to avoid damaging the delicate summer vegetation, already a subject of bitter complaint by the conservationists. Drilling started again on November 18 with the winter freeze. Progress was slow, due to washouts caused by the hot drilling mud melting the permafrost and the drill bit sticking on hard, fist-sized stones. It sometimes took several days to drill a couple of feet. Each time the bit had to be changed it meant pulling up the entire length of pipe, unscrewing it section by section, then going through the whole process in reverse to lower it again. Altogether, when completed at 12,000 feet, the well had used up eighty-seven bits. Forty might be a normal average.

It was mid-December by the time the permafrost layer was cleared and an outer casing run down to 2,000 feet. Once that was cemented in place, protecting the outer walls of the well, drilling speeded up to as much as 100 feet an hour. Cores were continuously taken to allow geologists to examine the various rock formations they were going through, and at 7,000 feet the tell-tale signs of natural gas were found, promising enough to make the first test. The result wasn't very encouraging. Gas flowed at a rate of just under one million cubic feet a day, reasonable enough for a well anywhere else in the United States but not worth developing in such a remote area as the North Slope.

At 8,600 feet Keasler carried out a second drill stem test. It was 1.55 am on December 27, 1967. This time the flow rate was stronger—so much so in fact that the needle on the 2,000 lb pressure gauge rocked right over on to its peg. Another gauge was fitted and with a tiny one-eighth of an inch choke registered a wellhead pressure of 3,075 lb. The gas was piped off to a pit some distance from the rig where it was flared. But because of a strong forty-mile-an-hour wind the flames were being carried dangerously close to the rig. Keasler didn't dare open the valve any further. Even so, the flow rate was one and a quarter-million cubic feet a day. Enough to give the six-man drilling crews, two working twelve-hour shifts while a third went on leave for three weeks in nine, a feeling that they might be on to something. Not that they knew the details. Security was all-important. Scouts—a polite name for industrial spies—would be sent by other companies to hang around the Fairbanks saloons to pick up any loose information from members of the resting drilling crew. Keasler didn't use the camp radio to transmit details to the company in case it was being monitored by the scouts. Each time he flew in specially to Fairbanks to call it through on the telephone.

It was after returning from such a call that the first serious problem arose. Because of the high pressure in the well, it was difficult to balance the weight of the mud column to hold back the gas. The well 'kicked' and the rubber packer became unseated. It was a tricky moment, with one of the crew standing by to close the blow-out preventer in case things got out of control. Eventually the gas was circulated out of the hole and vented off. But in shutting down the well it was found that the last 900 feet of drill pipe had become frozen solid at the bottom. Attempts were made to fish it out, after pulling up the rest of the pipe, but it was stuck fast. So the bottom section had to be plugged off with cement and a wedge lowered to 7,700 feet from where drilling side-tracked the original hole. All this took time—it was January 18 before drilling started again, when the opportunity was taken to enlarge the hole from nine and seven-eighths inches to twelve and a quarter inches. The general opinion at this time was that there was a possibility of a good find, but no one could be sure.

And so came the third test on the afternoon of February 18, 1968, between the depths of 8,750–8,883 feet. After shutting down the well for an hour it was opened up again at 3.18 pm for a continual-flow test of sixteen and a half hours, during which time there would be no rest for Jim Keasler or the geologist, Bill Penteller. Various choke sizes would be tried out, starting with a three-quarter inch. There were samples to be taken and analysed. And all the time they had to be sure the drill pipe did not become stuck again, as in the previous test.

Leading away from the rig was the pipe that would take the flow to a large pit, six feet deep and 200 yards square. The pipe ended a few feet over the edge. As the valve was opened up and the flow started, one of the men tossed a lighted torch into the pit. A huge fireball erupted with a roar, sending white and yellow flames fifty feet into the air. The colours signified natural gas. Keasler took a reading of the pressure to estimate the flow rate. He could hardly believe his own figures. It worked out at twenty-two million cubic feet a day. Penteller was taking samples and finding, as was to be expected, condensate liquids with the wet gas. These were crystal clear to begin with, but then they started to darken. They looked and smelled like crude oil. He turned to look for Keasler, but the engineer had walked across the snow to warm himself for a while by the fire of the flared gas. And it was there, half an hour after the test had begun, that he saw the fireball suddenly change to a deep orange colour. Thick black smoke rolled across the star-speckled sky. There was no doubt about it. They had gone through the lighter natural gas layer and struck oil.

'That was when we really began to get excited,' Keasler recalls. 'We felt we had made a major discovery—gas on top of oil, just the kind of structure our geophysicists had predicted might be there.'

He was right. They had located the biggest oilfield ever found in the United States.

The tests continued all that day. At one point Bill Penteller took his samples, sealed in one-gallon cans, to a heated shed. When he went to look at them later he found that because of the heat after the fierce cold they had twisted and expanded into the shape of footballs. In the early hours of the following morning Keasler flew in to Fairbanks to call the news through to Lee Wilson, Arco's district drilling superintendent in Anchorage. The fireball was visible from the plane for a hundred miles over the Brooks Range, lighting the Arctic darkness. Wilson had to get out of bed to answer the phone.

'Twenty-two million cubic feet?' His voice crackled over the phone with disbelief. 'You mean two million, don't you?'

'No—twenty-two,' Keasler insisted. 'And oil too.'

Wilson was only partly convinced, the figure was so much more than they could have hoped for. In fact, later more exact tests showed that Keasler had been conservative in his estimate. Forty million cubic feet a day was nearer the mark. For the moment, however, as Wilson passed the information to Jamison, who was now the company's manager in Alaska, and it went up through the chain of command, he attached a word of caution that more tests would be needed.

It was breakfast time in New York when the chairman and president of the company got to hear the news. A look at the geological map of Prudhoe Bay showed that of all the companies holding leases in that area, British Petroleum had the lion's share. A casual approach was made to BP, offering a substantial sum for its entire Prudhoe acreage. But BP wasn't to be fooled so easily. It had its own reasons for believing in the potential of Prudhoe Bay. Alwyne Thomas who had purchased the Prudhoe leases for BP when in charge of their Alaskan operations and was now regional manager in London had consistently recommended drilling at Prudhoe. As far back as 1966 he had suggested to Richfield that the two companies should drill a joint well there but the Americans had turned down the idea. Thomas now expressed himself firmly against Atlantic Richfield's proposal. As Dr Peter Kent, BP's chief geologist, commented: 'Our technical assessment of the potential did not encourage easy capitulation to our main competitor.'

Atlantic Richfield had already made a public announcement, two weeks after the second test, that the well had 'returned a substantial flow of gas'. The next announcement on March 7 merely stated that a second well was to be drilled in the Prudhoe Bay area, seven miles south-east of the first which was to continue drilling and testing. A week later, beset by press enquiries, the company revealed that in a test between 9,505–9,825 feet, the first well had flowed at a rate of 1,152 barrels of oil and 1·32 million cubit feet of gas a day through a three-quarter-inch choke. No mention was made of the much higher rate that Keasler had reported. It was not until June 25 that the company admitted that Prudhoe Bay State No. 1 had produced gas at forty million cubic feet a day and crude oil at up to 2,415 barrels a day, and at the same time that the second well, Sag River State No. 1, had found oil in the same formation and at the same level. Since it was seven miles away, all the signs pointed to the discovery of a major field. There were wild stories in Fairbanks of an Arco plane that had returned from Prudhoe Bay covered in oil after being caught in the spray of a gusher.

The company's shares rocketed on the stock exchange. More drilling would be necessary to find out the exact size of the field, but Atlantic Richfield themselves were pretty sure it was one of the biggest that had ever been discovered in the United States. This was confirmed a month later by a leading independent consultant, DeGoyler and MacNaughton of Dallas, who reported it as potentially one of the largest petroleum accumulations known in the world. The magical reserve figure of five to ten billion barrels of oil was mentioned for the first time, a fairly conservative estimate as it turned out, based on a recovery of only 40 per cent of the oil in the ground. What caught the imagination of the world was speculation that the entire North Slope might contain up to one hundred billion barrels of oil. It was the 'ton of gold' story all over again, except that this time it was oil. And it was from this point onwards, with large areas of the North Slope still available for competitive lease bidding, that security became as tight as any national defence project. No company wanted to let another know what it was doing or what it had found. The last chance had paid off and now the great oil rush was on.

7
The Need for Oil

At just about the time the Russians dropped their first official hints in Washington that they might be willing to sell Alaska to the United States another equally far-reaching event was taking place a few hundred miles away near a small country village in Pennsylvania. It was August 1859. The self-styled Baron Edouard de Stoeckel had returned from a visit to St Petersburg with new orders. Grand Duke Constantine, his ambitions fixed on furthering Russia's empire in Manchuria, had succeeded in convincing his older brother Tsar Alexander II that Alaska was not really worth keeping and should be sold to the United States. Stoeckel was instructed to find out if the American government, then under President James Buchanan, would be interested. Buchanan made an initial offer of five million dollars for the territory. That same month 'Colonel' Edwin Drake, also self-styled, created history of his own by discovering mineral oil in the ground at Titusville, Pennsylvania—the start of the modern oil industry that was to revolutionise the world. And while Stoeckel's efforts were delayed for eight years, first by the American Civil War and then by the haggling that took place over the price, Drake's discovery was an immediate success. Oil-rigs mushroomed all over the country as speculators and land-owners searched for this new source of wealth. Within eleven years John D. Rockefeller had founded the Standard Oil Trust that was to become the biggest private company in the world and for a while hold almost as much power as the government itself.

Petroleum was not new to man. Crude oil and natural gas seeping to the earth's surface had been known about and used for centuries. Where it became dried by wind and sun into bitumen it was used to reinforce mud bricks for building—by the Sumerians as far back as 3800 B.C., and by the builders of Babylon. The Bible records that Noah's Ark was caulked with bitumen. It was used by the North American Indians as a medicine, and the Mexicans found it made a good chewing gum. Natural gas escaping to the surface on the shores of the Caspian Sea, accidentally ignited in some remote age, became the famous 'eternal fires' which for 2,000 years were worshipped by religious sects. But what it was and where it came from was a mystery.

In more recent times the biggest demand for oil was as a fuel for lamps, but the source was originally whale oil. That was the reason for the big whale hunts of the early nineteenth century that brought both the Eskimos and the whales so close to extinction. Then in 1850 a Scottish chemist, Dr James Young, found that mineral oil could be distilled from coal and shale-stone. This product was used widely for a while, but the shale-oil industry could not compete after Drake's discovery that oil was to be found already in liquid form by drilling for it. Even that discovery was not really new, for there is evidence that the Chinese 2,000 years ago drilled for both oil and water. But like so many of the accomplishments of ancient civilisations, it was forgotten with the passing of time until there arose a need for it to be rediscovered.

What was new was the knowledge that by heating crude oil, and then condensing the vapours given off, it could be separated into liquids of different viscosity, from thin fluids to thick oils. The crude oil itself could be light or heavy, depending on the type found, but it was of no use for modern purposes until it had been refined by distillation. The immediate need was for kerosene as a substitute for the whale oil used in lamps, which with a growing population was in short supply. Other products left over from the simple distillation process were discarded as waste. But by an almost uncanny coincidence it was just at this time that inventors were trying to devise a portable engine that could drive a horseless carriage. Steam and electric power were tried at first, but they had serious drawbacks. What was needed was a light, easily transportable form of energy. And gasoline, one of the by-products from the new crude oil, was the ideal answer. It alone made possible the invention of the internal-combustion engine that was to change the face of the earth, giving man the automobile, wings with which to fly, and tractors to harvest his crops. Great industries arose from the energy that oil provided, ultimately sending man into space. But dominating them all, the source of their very lifeblood, was the oil industry itself.

hydrog... es
thousands of feet thick, comp... e.
Either in the form of oil or gas the petroleum ... gh
the fine pores of the stone. Where there was nothing to stop this migration it eventually seeped to the sea-bed and became dissipated in the water, or, if the sea by then had dried up, escaped into the air. These were the seepages known in ancient times. But if within the sedimentary stone there was a layer of clay, compressed into a watertight, non-porous rock, then the petroleum would accumulate underneath, with nowhere else to go. It is under such 'cap-rock' that oil- and gasfields are found.

Over the course of millions of years other geological changes were taking place. Mountains became worn away by the action of frost and rain, new ones were thrust up by upheavals under the earth's surface. Old seas and lagoons drained away to leave dry land. Continents split up and drifted apart, so that what is now the Arctic might once have been on the equator 250 million years ago. This is why petroleum can be found under land many hundreds of miles from the sea. Wherever it is now, the fact is that it was originally under a sea when it was formed. Other changes inside the earth caused the outer crust to wrinkle, forming hills and valleys on the surface and a similar undulating pattern in the layers of rock beneath. The ideal oil reservoir is a dome of cap-rock formed by one of these wrinkles, holding back the petroleum which has gathered in the sponge-like sedimentary stone underneath it. Natural gas, being lighter, collects above the oil layers. Under both is often to be found water-saturated stone. There are other types of reservoir, but the dome structure is the most usual. When the cap-rock is pierced by a drill—and it can be from a few hundred feet to many thousands of feet down, although 10,000 feet is a good average—petroleum is forced up by the pressure of the gas, rather like a soda syphon. This pressure can be very great. If uncontrolled and allowed to 'run wild' it can blow a drilling rig to pieces and send a column of flame hundreds of feet into the air which would burn for perhaps twenty years unless cut off. In seeking to extract petroleum from beneath the ground, man is tapping one of the most powerful forces of nature. This was little understood in the early days of drilling, resulting in the 'gushers' of oil that were produced. Even today blow-outs sometimes occur which cause loss of life and oil-well fires requiring many weeks to put out. Drilling for oil is a dangerous business.

Drake's discovery led to an immediate oil boom. The demand was already there, since mineral oil was cheaper than whale oil. And big profits were to be made, for under American law oil belonged to the person owning or leasing the land on which it was found. This is so today in the United States, whereas in nearly every other country in the world oil rights belong to the government. The most a private owner can usually expect is a fixed rental from the oil company operating on his land while the royalties go to the government in the form of taxes. But matters were complicated by another American doctrine which held that even though a well drilled on one plot of land might be draining oil from a reservoir extending under several adjoining plots, the oil still belonged to the owner of the well extracting it. Because of this 'law of capture' it was in the interest of an owner to extract every drop of oil as quickly as possible before the field became depleted, just as it was in the interest of others to drill immediately on the neighbouring plots. Oil-rigs sprouted like forests, prematurely exhausting the oil-fields, for to get the most out of any reservoir it should be produced at a steady and economic rate. Such inefficient methods also had the effect of causing a glut of oil in the early days, with prices falling as competition became fiercer. It was not until the 1930s that the oil-producing states, notably Texas, introduced 'prorationing' regulations to limit the number of wells that could be drilled in a given area and also the amount of oil that could be produced, depending on a field's capacity and current demand. This had the advantage of conserving the nation's oil, but conversely, used as an instrument for maintaining prices which have consistently been higher than anywhere else in the world, it has led to continued production from thousands of small producing wells which would otherwise be uneconomic.

In the late nineteenth century, however, before prorationing, the American oil industry fell into a state of violent disorder and it remained so until it came to be controlled almost entirely by one man, John D. Rockefeller. He cunningly began not by joining the rush to produce oil but by taking over the refineries which were vital in processing the products required. His Standard Oil 'Trust' was eventually declared illegal by the Ohio Supreme Court when it threatened to hold the whole country to ransom and was dissolved in 1899. Again in 1911 the US

Supreme Court held that the Standard Oil Company of New Jersey was violating the Sherman anti-trust laws and compelled the company to distribute its holdings amongst subsidiaries, such as Standard of Indiana and Standard of Ohio (the original unit on which the trust was formed). From then onwards these companies operated as separate entities with no connection with one another, other than their names in some cases. But the robber-baron image of the original trust has never been completely forgotten. It remains an albatross round the neck of the American oil industry to this day, whenever criticism of its methods is aroused.

Demand grew rapidly as new uses for oil were discovered. Gasoline for automobiles quickly took over from kerosene as the main product required. Oils and greases were needed for lubrication and, with the First World War, most navies of the world began changing from coal to fuel oil as the source of power for their ships. Other countries clamoured for oil products, but it was soon found that through one of the ironies of nature, oil usually existed a long way from where it was most needed, in undeveloped countries that had little use for it themselves. The only two exceptions to this were the United States and Russia. There was a small production in Rumania around the time of the American discovery, but in general comparatively few oil or gas-fields were found in the industrialised countries of Europe or Japan. This remained so until the important discoveries of the 1960s in Holland and offshore in the North Sea.

It was only the United States and Russia that possessed both large reservoirs of their own oil and large domestic markets to benefit from these. Russia, by 1900, had an output exceeding that of the US, but the private owners were expropriated after the revolution and in government hands the industry declined. There was little oil available for export after domestic needs had been met. It is only recently, since the mid-1950s, that exports have been resumed on any scale. In the United States, on the other hand, the industry forged ahead after the first years of uncertainty. It was primarily on the ready availability of petroleum that America's twentieth-century prosperity was based, a home source of energy where other nations had to import theirs. And the country they had to buy it from was the United States. Since the oil was originally shipped over in barrels, this came to be the form of measurement generally used in the industry, one barrel being equivalent to forty-two US gallons and, depending on the specific gravity of the type of oil, seven and a half barrels being equal to one metric ton.

Of the numerous oil companies formed in the United States, the largest and most powerful were the 'big five'—Jersey Standard,

Texaco, Gulf, Standard of California, and Mobil (originally known as Socony-Vacuum). These were involved in every stage of the industry, from producing oil to refining, transportation and marketing. They dominated not only the American market but also, through their exports, the oil-bereft countries of Europe, to the point where it caused serious political concern. Europe desperately wanted its own sources of oil so as not to be too dependent on the United States for it. If it was not available indigenously, then the only other answer was to look for it abroad. A number of companies backed by European capital and sometimes by governments themselves were formed for this purpose but only two became big and successful enough to challenge the American giants, eventually standing equally with the big five to make up what came to be known as the major international oil companies. These were Royal Dutch Shell and British Petroleum (originally called the Anglo-Persian Oil Company). They pioneered the search for oil in the more remote parts of the world where large sedimentary basins were known to exist—primarily in Venezuela and Indonesia in Shell's case and the Middle East in BP's, where the first discovery was made in Persia in 1908. With the later discovery of huge oilfields in Iraq, Kuwait and Saudi Arabia, the Middle East became the key not only to supplying the oil-hungry nations but also to international politics. It was the most prolific oil-producing region in the world, containing two-thirds of the world's proved reserves. Its oil was also much cheaper to produce than that of the United States—about ten cents a barrel compared with over a dollar. So marked was this difference that the American majors soon wanted a share of the action and extended their own operations overseas, both in production and marketing.

From the 1920s onwards, the international oil industry was dominated by the seven major companies—the five American, one British, and one Anglo-Dutch—of which Jersey Standard and Shell were the largest. Between them the 'seven sisters', as they were caustically dubbed by less successful rivals, virtually controlled all phases of the industry from production to marketing in most countries of the world, excluding the Soviet bloc. This has been less pronounced since the mid-1950s. For one thing there has been a considerably growth in the international activities of other integrated oil companies such as Continental, Phillips, Union of California, Atlantic Richfield, Getty Oil and Cities Service in the United States, the French Compagnie Française des Petroles, and Petrofina of Belgium. Also, some nations have formed state-controlled organisations both to produce and buy oil for their own markets, dealing on occasion direct with producer countries who have set up their own state agencies to sell oil. However, the seven majors still account for more than half of the oil production in the non-Communist world. Their gross revenues in 1970 were nearly sixty billion dollars. They own or charter most of the world's tanker fleet, which itself accounts for over a third of total merchant shipping. More than half of the world's cargo being moved at any one time is petroleum.

Until the Second World War it was the practice of the major companies to build refineries close to their oilfields, even when these were far away from the industrial markets, and then transport the products by ship, pipeline and rail to the consumer. The great Abadan Refinery at the head of the Persian Gulf, built by BP to process Iranian oil, was a typical example. But after the war this pattern changed as companies began to favour a policy of siting refineries close to the consuming markets instead. This had the advantage that it was simpler to ship crude oil in bulk rather than make the same journey with different products that had to be kept separated and might require different conditions of transportation. For oil was becoming vastly more complex. Scientists had found many more uses for it in the chemical and plastic industries. The most important factor, however, was that the refineries, representing enormous capital outlays, were more secure on home ground where they were not vulnerable to appropriation by foreign governments. International politics had gone through yet another turnabout. In seeking their sources of supply from the Middle East, the industrialised countries had originally found those governments only too keen to permit exploration and production of oil in areas which were mostly barren desert. Where successful, it meant large sums of money being paid to them in royalties with absolutely no effort or risk on their part. But this sudden income to previously impoverished Arab nations caused social and political upheavals and the emergence of an aggressive nationalism that demanded an ever-increasing percentage of the oil company's profits. It came to a head in 1951 when the Iranian government took over by nationalisation the Abadan refinery and BP's oilfields in Iran. It was a wild gesture and in the event the Iranians found they could not operate the industry themselves. After a few years they had to call the oil companies back. But from that moment onwards the countries of Western Europe and the Far East could no longer count on their sources being guaranteed secure. They were uneasily dependent on the whims and caprices of Middle Eastern politics which on two occasions, during the Arab-Israeli conflicts of 1956 and 1967, caused a complete stoppage of supplies and a shutdown of the Suez Canal which is still in effect.

In both those crises the United States was able to come to the aid of the free world by stepping up its own production and that of

Venezuela. But the situation in the US itself had changed. In the years following the Second World War oil had been imported mainly because it was cheaper than that produced domestically. Towards the end of the 1950s, however, Americans suddenly woke up to the fact that they were beginning to run out of oil. A fast-growing home demand and a fall-off in new discoveries meant that the nation, for the first time in its oil history, was having to rely on imports from abroad to make up the difference. This is still the trend. Demand in the US is by far the highest in the world, equal to over 1,100 gallons of products a year for every man, woman and child. The comparable figure in Britain is just under half of that amount. In Russia it is about 350 gallons, and the average for the world is less than 200 gallons. The US consumes some 30 per cent of the world's total energy, although with less than 7 per cent of the world's population.

The world demand for oil has just about doubled every ten years since the industry started and proved reserves at any one time—that is, oil that has been discovered and can be produced economically by existing techniques—have never allowed for a foreseeable supply of much more than twenty years ahead. There have been numerous gloomy predictions since the turn of the century that the world will run out of oil, but new sources have always been found to meet the demand. The cost of finding it, however, becomes greater each year as the more accessible fields are depleted. In 1969 proved world reserves totalled 525 billion barrels while world consumption was sixteen billion barrels. This consumption is expected to double over the next ten years, at which rate of growth reserves would be exhausted in less than twenty years. To maintain the same reserve to production ratio, 900 billion barrels of new oil must be discovered by 1985, more than the whole amount that has been found since the industry first began. That is the problem. Other sources of power such as nuclear energy will become increasingly available, but by 1985 oil and natural gas will still have to supply 60 per cent of the world's total energy demand. Much of the new oil required could undoubtedly be discovered in the Middle East, which still contains about two-thirds of total world reserves. But political instability in that area and the ever-increasing financial demands of the Arab governments have encouraged the oil companies to extend their search elsewhere. This led to the big discoveries in Nigeria and Libya in the 1960s and the growth in offshore exploration beneath the sea-bed of the continental shelf in many parts of the world. Offshore production accounts for about 16 per cent of the present world total and will be even more in future. Failing the discovery of enough new oilfields, there is always the possibility of producing oil from shale-stone, tar, sands, or even coal, on which a great deal of scientific research has been carried out in recent years. Although uneconomic at today's price levels, these sources are great enough to keep the world supplied for several hundreds of years if necessary.

As an oil-producing region the Middle East is the most important in the world, but the biggest single producing country is still the United States. Its current yearly production of over four billion barrels is some 24 per cent of total world production. This has meant eating increasingly into the country's oil reserves, which at the beginning of 1969 stood at little more than eight years' supply at those rates. With demand running at 5·5 billion barrels a year, the US now has to import about 25 per cent of its oil requirements. Most of this comes from Venezuela and Canada, with only 4 per cent from the Middle East. Domestic production could be increased in the event of an emergency by opening up wells to their full capacity, but the amount thus available grows less each year. By world standards, many of the producing wells are inefficient and uneconomic. Two-thirds of the nation's 573,000 oil-wells are marginal, averaging only 3·6 barrels a day each, whereas a single Middle East well might produce 20,000 barrels a day. Because of high costs, new discoveries have not kept pace with demand. And that is the importance of Alaska. At one stroke, if the more optimistic estimates are correct, the Prudhoe Bay oilfield may have doubled present US reserves of thirty billion barrels. If, as may be possible, further discoveries of a similar nature are made on the North Slope and elsewhere in the state, Alaska will become the chief oil-supply region for the United States, taking over the role that Texas held for so long. If the discoveries are big enough and the North-west Passage can be opened up for regular shipping, northern Alaska and the Canadian Arctic, uniquely situated midway between the markets of the Far East and Western Europe, could become a major oil-exporting region. The Arctic as a whole, where in Siberia the Russians have already discovered huge deposits of oil and natural gas, including the world's largest known gas field estimated to contain at least 200 trillion cubic feet, could become the richest petroleum province on earth.

8

Why Alaska?

In ranging round the world in the search for oil the first clue an oil company looks for is a thick underground sedimentary basin of porous sandstone or limestone, for it is only in this kind of formation that petroleum can be formed. These basins, sometimes many thousands of square miles in area and extending under whole countries and seas, have been well-defined in most parts of the world. The question facing the geologist and geophysicist, who are the vanguard of any exploration effort, is which of them might contain petroleum, and where.

Ultimately, even with all the help of modern science, the only way to answer that question for sure is to drill a well, and maybe a whole series of wells if the first is a failure. The success ratio is not very high. On a world basis, only about one wildcat exploration well in twenty makes a worthwhile discovery. In New Guinea, for instance, companies have been searching on and off for thirty years and spent around forty million pounds without finding a drop of oil. Most of the shallow, more easily located oilfields have already been developed, so that problem becomes increasingly more difficult. Drilling for oil is always a gamble, with no guarantee of getting back any of the large investment required.

The ideal location on which to drill a well would be something like this: a sedimentary basin containing plenty of marine shales, as revealed by fossils, with intervals of cap-rock to seal the porous formations and the layers of rock folded gently and not violently by earth movement so that they are visible at the surface. There should preferably be seepages of oil or gas in the vicinity to indicate the presence of petroleum, and the area should be in a temperate climate close to a large industrial market—or, if not, at least near to deep water to facilitate transportation by tanker. In addition, the government of the country should be politically stable. It is rare for all of these conditions to be met. The choice depends on how far they fall short of the ideal in any particular respect. And even where perfection seems to exist it might be found by drilling that petroleum had indeed been present but had seeped away through cracks in the cap-rock.

What are the factors that influence an oil company in its decision to explore in a particular area? The most obvious is if another company has already found petroleum there. In such an event others will rush to buy up leases on the adjoining land, particularly if the strike is a big one, for the chances of finding more petroleum are that much greater. This is why the attention of the industry can switch so abruptly from one area to another—to the North Sea after the natural gas finds in the mid-1960s, for instance, or to the Arctic as at present. If the company is considering a new and previously unexplored area, the first essential is for the country concerned to have adequate mineral and mining laws so that land can be leased without undue complication and it is known what kind of taxes and royalties will have to be paid. There could be great problems if oil was discovered on a disputed boundary between two countries, or if there were conflicting claims to land ownership. Technology is another factor. It is no good locating oil if there is no way of extracting it. Offshore exploration is a case in point. It only became possible with the development of techniques for drilling at sea, with operating depths increasing as equipment and methods improved. And the price for which any oil discovered can be sold is a vital consideration. A small field in the Middle East might not be worth producing economically at the prevailing posted prices, whereas it would in the United States where prices are more than double. The reasons for this are higher production costs in the US and the fact that in order to protect the domestic industry against foreign competition and so maintain secure supplies, the United States has since 1959 imposed restrictions to limit the importation of cheaper oil from overseas.

Various scientific surveys are made of possible oil-producing areas. The first is usually an aerial reconnaissance to study the general surface structure which also gives a clue to the kind of formations that are underground. Then geologists examine the rocks and outcroppings to assess their age and fossil content, to see whether they are the right type to contain petroleum. If an area continues to look promising, geophysical surveys are an essential prelude to any drilling. The most

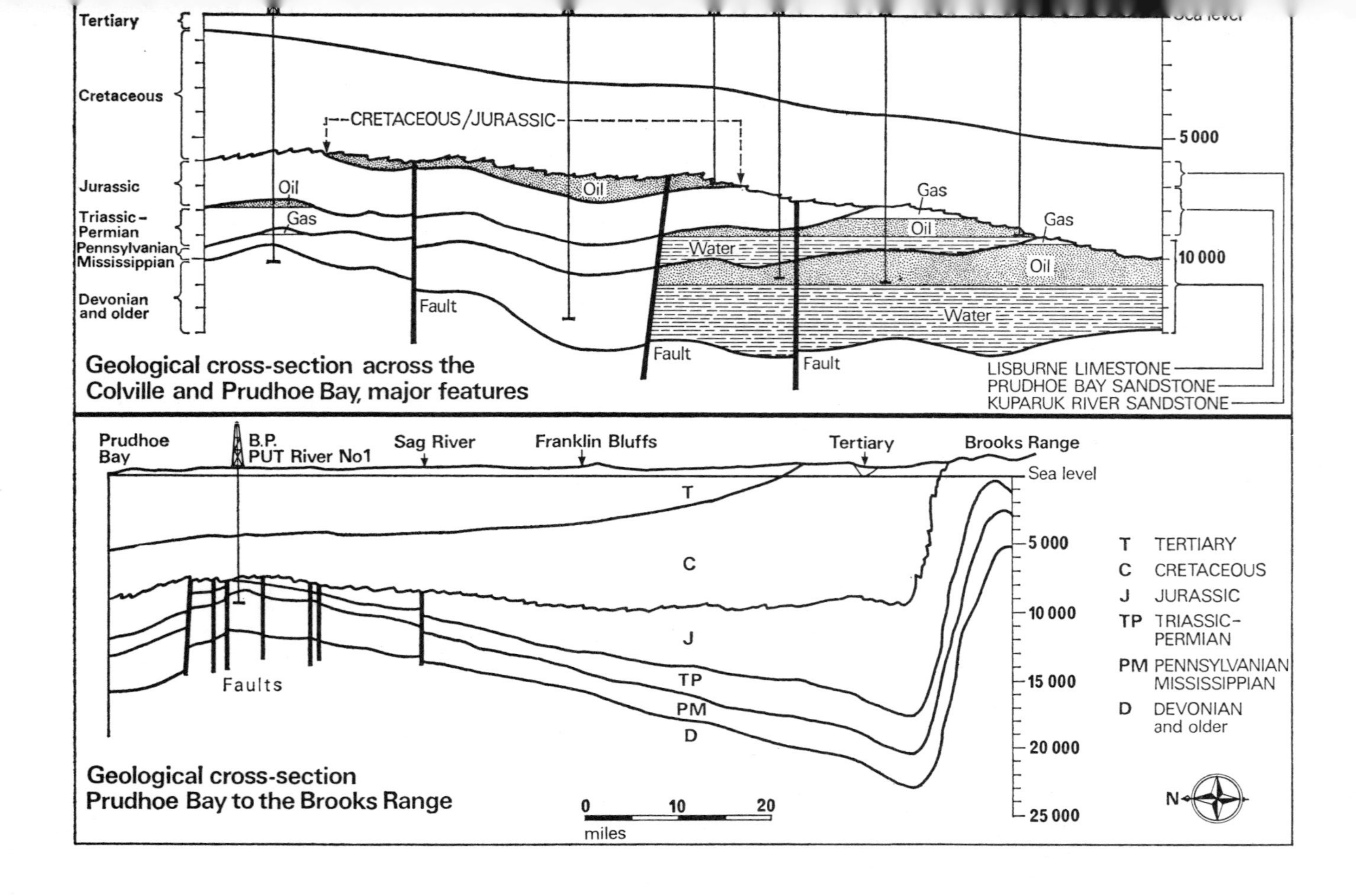

Geological cross-section across the Colville and Prudhoe Bay, major features

Geological cross-section Prudhoe Bay to the Brooks Range

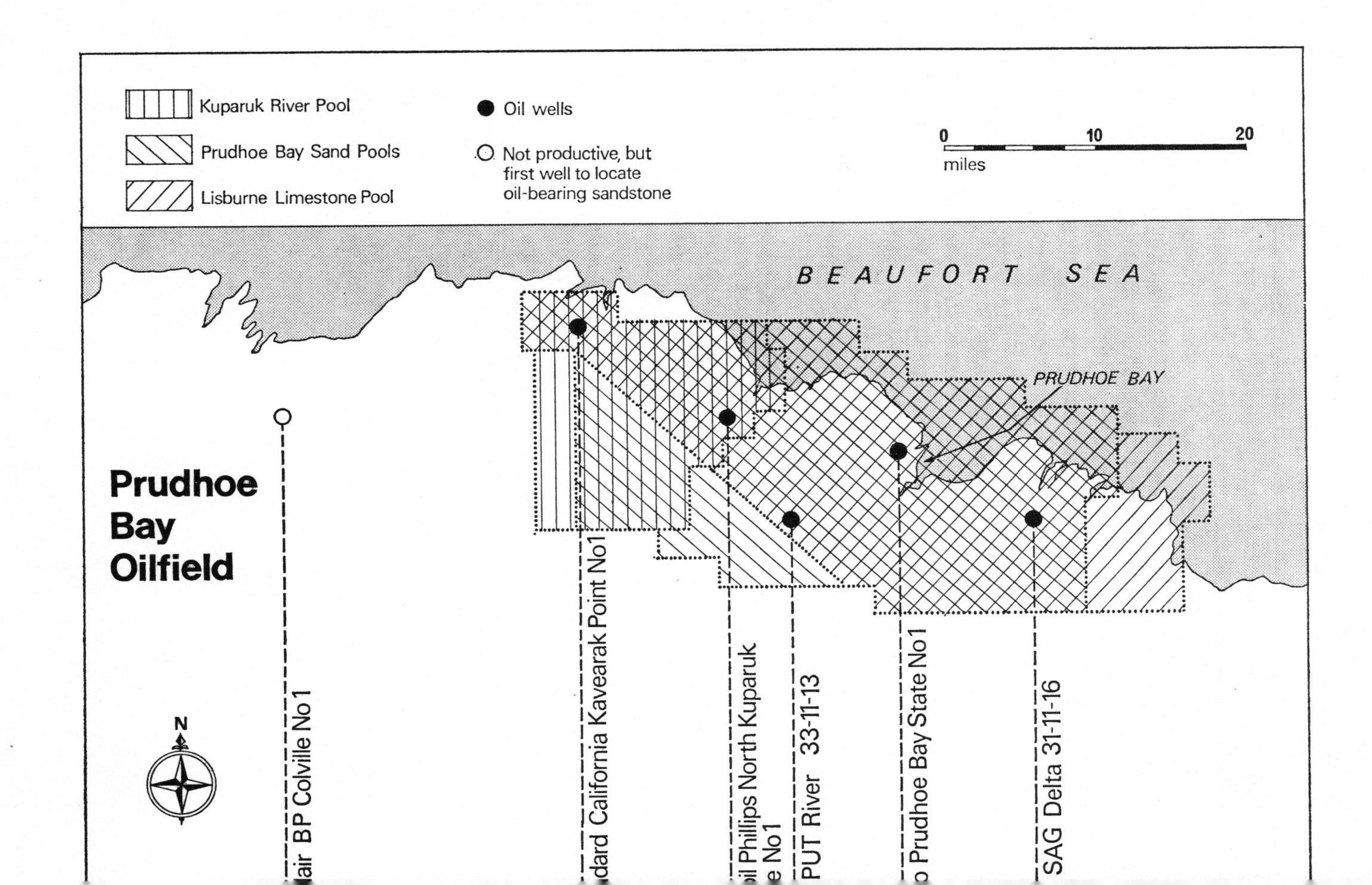
Prudhoe
Bay
Oilfield
Kuparuk River Pool
Prudhoe Bay Sand Pools
Lisburne Limestone Pool
Oil wells
Not productive, but first well to locate oil-bearing sandstone
0
10
20
miles
BEAUFORT SEA
PRUDHOE BAY
N
air BP Colville No 1
dard California Kavearak Point No1
il Phillips North Kuparuk
e No1
PUT River 33-11-13
o Prudhoe Bay State No1
SAG Delta 31-11-16

widely used is the seismic survey, in which small charges are exploded at regular intervals along a line on the surface and a record made of the time taken for shock waves to travel down and back again from the sub-surface rock layers. These vary according to the different types of rock and their depths. In this way a detailed picture can be built up of the various formations, revealing the presence of any dome-like structures —anticlines, as they are professionally called—under which petroleum may be trapped. Magnetic and gravity surveys are also used to measure the magnetic properties and density of rock formations to indicate where they might have been located in the past, for it is generally accepted today that the continents have drifted in geological time over the earth's surface. This is a vital point. It is known that oil can only be formed in marine sediments in tropical conditions, just as certain corals can only grow in warm water. Most of the large oil-producing areas of the world are in fact within the tropics, formed in recently deposited rocks—recent in a geological sense, that is, meaning about thirty million years, less than 1 per cent of the earth's life. Oil found in higher northern or southern latitudes is contained in older formations which have moved in the course of time. Alaska might not at first appear to be a likely place to find oil. But its sedimentary rock—there are thirteen known basins in the state, one of which underlies the entire North Slope—was laid down in an earlier age, perhaps 250 million years ago. It was then close to the equator (as shown by the fossil evidence of tropical vegetation) and the conditions did exist for the accumulation of petroleum. This was first revealed by the tell-tale signs of natural seepages, which have always been such a useful indication of petroleum reservoirs.

The Russians first discovered oil seepages in the vicinity of Chinitna Bay on the Alaska Peninsula soon after the beginning of the nineteenth century, although the presence of this phenomenon had long been known to the Indians. The Russians did have some interest at one time in extracting the seepages—an oil trade based on similar seepages in the Baku region of the Caucasus had already been in existence for many years. But demand for the bitumen-like oil was limited mainly to its use for making bricks and Alaska was too far away from any possible market for the costs involved to make it worth while. It was not until 1892, after Alaska had been bought by the United States and when the modern oil industry was already well established, that claims were staked in the area with a view to drilling for oil. Six wells were drilled between 1902 and 1906 by the Alaska Petroleum Company and the Alaska Oil Company, but they were unsuccessful.

Meanwhile, even more extensive seepages along the shores of the

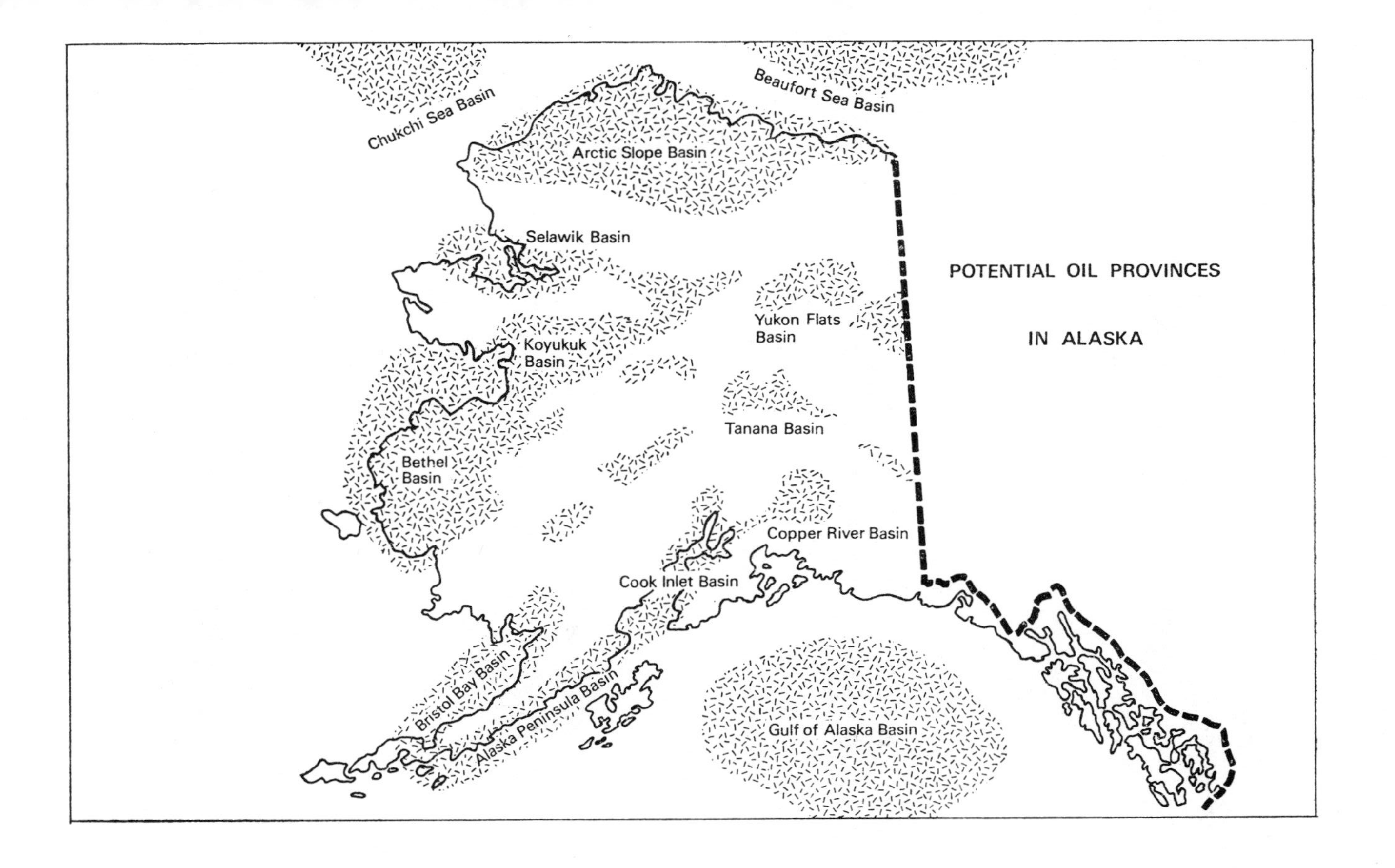
POTENTIAL OIL PROVINCES
IN ALASKA
Chukchi Sea Basin
Beaufort Sea Basin
Arctic Slope Basin
Selawik Basin
Koyukuk Basin
Yukon Flats Basin
Tanana Basin
Bethel Basin
Copper River Basin
Cook Inlet Basin
Bristol Bay Basin
Alaska Peninsula Basin
Gulf of Alaska Basin

Gulf of Alaska from Cordova to Cape Spencer had been noted by miners coming northwards in the great gold rush, and some of these men turned their attention to the possibilities of petroleum instead of gold. Claims were staked in 1896 at various places along the coast, resulting in the drilling of a well in 1901 at Katalla, forty miles southeast of Cordova, which the following year was brought in as Alaska's first oil-producer. One reason for this early interest in oil was to supply kerosene for lighting and heating to the many prospectors who were flocking to the territory and which was expensively having to be shipped in by the Standard Oil Company marketers. The Chilkat Oil Company, formed to produce the oil, drilled eighteen wells in the Katalla field up until 1932, the deepest being 1,800 feet. The total of 154,000 barrels produced was processed in a primitive refinery built nearby and the products sold locally. In 1933 this refinery was destroyed in a fire, however, and it wasn't considered economical to rebuild. The whole operation came to an end. Further drilling was carried out in 1926 by Mobile Oil (through the General Petroleum Corp) near Yakataga, 150 miles to the east, but although some shows of oil and gas were encountered, the well was abandoned.

Interest returned to the Cook Inlet area on the Alaska Peninsula, where at various times between 1923 and 1939 a number of deep test wells were drilled in the Kanatak and Chinitna districts and at Chickaloon in the Matanuska Valley. The main companies concerned in this exploration were Standard Oil of California, Tidewater, and Union Oil. Oil shows were discovered, but not in sufficient quantity to be worth producing. Although the main drilling activity had been confined to the southern coast of Alaska, the most promising area from a geological point of view seemed to be in the Arctic, where Alaska's largest sedimentary basin underlay the entire North Slope beyond the Brooks Range. At that time, however, it was too costly for private industry to consider exploring there. Instead, it was the US Navy who pioneered the way by proving the potential of this region.

The history of the North Slope shows that it might well have become British territory had Captain Cook and Sir John Franklin thought of planting a few Union Jacks on that windblown wilderness of ice and snow. Captain Cook was the first white man to see it when in 1778 he sailed northwards through the Bering Strait as far as Icy Cape at the western end of the Slope. Apart from the Eskimos—and the empire-builders of those days certainly never confused native occupancy with legal ownership—no one else had ever been there. The Russians vaguely described it as a part of North America and left it at that. But the British were not very interested either, happily coming to an agreement with the Russians in 1825 to establish the boundary between their respective North American possessions at the 141st parallel of longitude, an imaginary line which at its Arctic extremity meant next to nothing to the signatories beyond seeming to be a convenient point on the map. Britain's main concern at that time was in finding a north-west trade route between the Atlantic and Pacific oceans. This was the reason for the next white visitors in 1826 when two ships of the Royal Navy attempted to make the passage by approaching separately from different directions—Sir John Franklin from the east and Captain Beechey from the west. Both were stopped by ice within 150 miles of one another, Franklin reaching the Colville River delta and Beechey the most northern promontory which he named Point Barrow. Forced to turn about and go back the way they had come, both men took the opportunity of charting the coast and naming various of the land features, mostly after the names of their crews. Over succeeding years, especially during the abortive attempts between 1848 and 1853 to try to find out what had happened to Franklin and his crew after they had failed to return from a voyage, this work was completed by other British ships and by pioneers of the Hudson's Bay Company.

American interest in the region began in 1848 when a whaler entered Arctic waters by way of the Bering Strait, leading to the development of a substantial whaling industry. Other nations also took part in the dangerous but profitable search for whales, bringing about the first contact on any scale between the Eskimos and white men. It was disastrous from the Eskimo point of view. Not only were they deprived of a major means of subsistence, but their numbers were severely reduced by the white man's diseases—tuberculosis, influenza and venereal disease—against which they had little resistance. Some of the whale-hunters came ashore to settle with the Eskimos, one in particular being Charles Brower, of Dutch extraction, who took to trading instead of whaling and became famed as 'King of the Arctic'. At some time around 1890 he reported to the US government the existence of oil seepages on Cape Simpson, sixty miles east of Barrow, which the Eskimos had for many years been using with driftwood to make fires for melting down whale blubber. No notice was taken of the report and it was forgotten until 1917 when the seepages were rediscovered by a visiting official, Alexander M. Smith. By this time, the oil age was fast approaching. The seepages were large enough to warrant investigation by the US Geological Survey, who confirmed the presence of oil-bearing formations. The Standard Oil Company of California sent one of its own geologists there in 1921 and actually staked out a claim—the first in that region. But the climate and terrain were so inhospitable and the problems of drilling so formidable that nothing more was done about it.

The First World War had shown that the US Navy would require immense quantities of oil now that ships were converting over from coal—as indeed would all the navies of the major powers. It was of national concern to be sure of a secure source of supply without having to rely on the commercial market. In Britain this took the form of a government decision in 1914 to invest in the Anglo-Persian Oil Company (later BP) as majority shareholder—nationalisation under another guise, in fact—and to enter into a long-term contract for the supply of fuel oil to the Royal Navy. Another method was adopted in the United States. The government selected large areas of land which were likely to contain oil and set them aside as naval petroleum reserves, on which no private leasing or drilling was allowed. It was not the intention to exploit them immediately, only to have them available in case of emergency or if supplies failed from other sources. In 1923 such a reserve was established in Alaska covering the western half of the North Slope, twenty-three million acres around Point Barrow where seepages had been observed. This became Naval Petroleum Reserve No. 4—an area as large as the state of Indiana—which more than any other single factor was responsible for the eventual discoveries further along the North Slope. It was that same year on similar federal oil land in Wyoming known as the Teapot Dome that Secretary of the Interior Albert Fall caused a political scandal by leasing rights to private industry, the repercussions of which linger to this day.

Geological surveys were carried out in the reserve from 1923 to 1926, but it was not until 1944, with the shortage of oil created by the Second World War, that the federal government embarked on large-scale exploration there. Over the following ten years, until the programme was discontinued in 1953, thirty-six test wells and forty-four core holes were drilled at a total cost of sixty million dollars. The main result of the work, undertaken by the US Navy and the Geological Survey with the help of other federal agencies and private contractors, was the discovery of three oilfields and six gasfields, none of the commercial propositions. The biggest oil find was at Umiat on the Colville River near the eastern boundary, where reserves were estimated at seventy million barrels. The major gas discovery was at Gubik, with an estimated reserve of twenty-two billion cubic feet in tested sands and possibly 300 billion if untested sands of similar characteristics were included. It was a smaller gasfield however, discovered near Barrow in 1949, that proved the most useful, supplying power for the base camp and now being used as a source of light and heating in Barrow and the nearby Naval Arctic Research Laboratory. These were not major discoveries, although exploration came to an end before their full potential could be tested, but they did point to the probability of large petroleum accumulations in the region.

The Navy project ended the first era of oil exploration in Alaska. It had met with a marked lack of success. Over a period of fifty years something approaching eighty million dollars had been spent on the search by the government and oil industry for the actual production of only 154,000 barrels of oil—from the old Katalla field which in any case had ceased operating in 1933. Since then not a single drop of oil had been produced commercially in Alaska and natural gas was supplying only one local Eskimo village and a naval research establishment. It looked as if Alaska would remain both non-producing and also a mere territory, for although much attention was now being directed towards the fight to achieve the status of statehood, it still looked like being a long way off. With its relatively tiny population and lack of economic viability, Alaska was almost totally dependent on federal grants and development money, in spite of the great contribution that its resources had made and were still making to the nation's wealth through outside business concerns. The seeming fact that Alaska could not stand on its own feet was one of the main arguments used by those opposing statehood.

But in a remarkable turn of fortune, which in fact is not all that unusual in oil exploration, both these factors changed overnight. A major oil discovery on the Kenai Peninsula in 1957 led to Alaska becoming an oil-producing region—within a few years ranking eighth in the nation, even apart from the North Slope—and was the economic stimulus for the granting of statehood two years later. Alaska would undoubtedly have been made a state eventually, but it was oil that tipped the balance at that particular time. It also marked the beginning of apparently unrelated problems to do with conservation, native claims, and land ownership. For as chance would have it, that first big discovery was made right in the middle of a moose reserve. It awakened the first murmurings of criticism from environmentalists. It gave the leaders of Alaska's native groups something tangible to fight for. And it also created the first of the new Alaskan oil tycoons.

9

Moose, Money and the American Dream

Locke Jacobs came to Alaska to make his fortune. He was not alone in that. Many people came to Alaska for the same reason. It has long been part of the great American dream for a young man to venture into the frontier regions to make good. It just happens that Alaska is the last frontier where it can still be done in the old pioneering way. For most, including those who had a lot more going for them than Jacobs, the dream faded with the harsh reality of life. They settled for less and were probably quite content for it. But not Jacobs. He arrived in Alaska in 1946 at the age of twenty-two with just twenty-eight dollars in his pocket—$100 given to him by his father on leaving home in the timber country of Oregon had been lost in a friendly card game on the way over. Today his yearly income is around $900,000. He is a millionaire many times over. He made most of it buying and selling oil leases.

Anchorage in 1950 was a bustling little town of about 40,000 people. One of the leaders of the business community was Robert B. Atwood, a newspaperman from Springfield, Illinois, who had come up in the 1930s to take over a local newspaper, the *Anchorage Daily Times*, which then had a circulation of 650 copies a day. He had built this up more than ten-fold and was now in the forefront of the fight to achieve statehood. Like other prominent citizens who were concerned with the Alaskan cause, including his father-in-law Elmer Rasmusen, president of the Bank of Alaska, he was worried that a large part of Anchorage's economy was dependent on the presence of the US military who had remained after the war to keep a watchful eye on the Russian threat from across the Bering Sea. If the military ever decided to leave, Alaska would sink back again into the depression from which it had suffered so much before. New industry was needed to balance the economy and also to show that Alaska could stand on its own feet if it was ever to become a state.

'Ever since the early forties the oil companies had been sending field parties up here,' Atwood recalls. 'They'd come to Anchorage to pick up camping equipment—no helicopters or anything like that in those days—then disappear into the boondocks all summer. We'd just see them in the spring and then again in the fall before they'd go back to California. They kept saying that the Kenai Peninsula looked good for oil, but they never seemed to do anything about it.'

Bob Atwood was speaking in the plushly modern office of his newspaper, which now sells over 40,000 copies a day. He has become a prosperous publisher, still briskly energetic, although with no longer a cause like statehood to bring the light of battle to his eyes. He speaks wistfully of those days—and the disillusionments that followed after the battle had been won. He thinks of retiring, but 'you get to wondering what you're going to do after breakfast'. As a leader of the 'establishment' in Anchorage, he is pressing hard for the development of North Slope oil against the blocking tactics of the conservationists. 'This is the golden age of kookery. Everyone's sounding off. If they think the companies are going to bury a hot pipeline in permafrost and chance losing the oil, they're crazy. The companies have a common interest with proper conservationists. But these other kooks—they talk wildly and people seem to listen to them. Even they'd go hungry if there wasn't a certain amount of pollution. And if we don't have oil—what then? We'd have to cut down the trees for fuel. Is that what they want?'

Atwood had just come back from a visit to the North Slope. Much controversy was raging over accusations that the oil companies had littered the area with garbage. 'The only debris I saw was a pile of film wrappings where photographers had been out on the ice taking pictures of those apple trees.' The trees in question were made of plastic, the gimcrack idea of someone in BP's head office in London. They had been brought all the way over from England and 'planted' on the Arctic ice so that a photographer could take pictures of them for an advertising campaign. At a time when pollution and environmental damage were such thorny questions they were a great embarrassment to local BP officials and the cause of much hilarity amongst the oilmen on the North Slope. Standing in their proud plastic greenery amidst hundreds of miles of featureless Arctic ice, they had become a kind of tourist attraction, a symbol of big-city silliness.

But all that came later. Fifteen years before, there were no oil companies. The problem was moose.

By the early 1950s, Atwood and his friends at the local 'Spit and Argue' businessmen's club had become exasperated at the failure of the oil companies at least to test by drilling what seemed a good oil prospect on the Kenai Peninsula. They knew nothing about the industry, but every once in a while they'd see leases filed by outsiders at the Bureau of Land Management. About thirty of them got together and decided to take out some leases themselves and then try to get someone to drill a well. A friendly geologist told them that they had to have a block of at least 60,000 acres in order to get an oil company interested. They each put in what they could afford and then looked around for someone who understood the mysteries of leasing and who could handle it for them. Enter Locke Jacobs.

Jacobs had been bumming around Alaska in traditional style, looking for the right chance. He worked as a section hand with the railroad for a while, then got a job as steward on board the *Nenana*, last of the old Yukon sternwheelers which now rests in Fairbanks as a museum-piece. He had been interested in geology since college days, and with a few dollars saved up he bought some books on the subject which he studied during the journeys up and down the Yukon. Gold prospecting would have been an obvious choice, but he was more interested in the possibilities of other minerals which had been to a large extent neglected. Staked by the captain of the river boat, who happened to have a degree in geology and was impressed by his ambitious young steward, Jacobs set off into the hills to look for minerals. After some time spent in roving round the territory he decided there wasn't an immediate future in mining—although he now believes it will boom in about five years' time. He came to Anchorage, where the money was, and after a number of odd jobs finally went to work as a stockroom clerk in an Army-Navy surplus store. A friend had loaned him a book on *Practical Oil Geology*—he still has it in fact; when the friend asked twenty years later when he was going to return it Jacobs reminded him that he had only promised to return the book when he was finished with it and he had not yet reached that point. Armed with scant knowledge but a lot of enthusiasm, Jacobs turned his attention to the oil business. He bought some maps from the office of the US Geological Survey and with $170 of savings invested in an oil lease at Iniskin Bay on the Alaska Peninsula.

Oil leasing has long been a 'little man's gamble' in the United States, a democratic gesture which in theory at least gives him the chance of a stake in the natural wealth of the country. Ever since the Mineral Leasing Act of 1920 anyone, whether a large company or an individual person, can acquire an oil lease on federal land for the payment of fifty cents an acre—it was twenty-five cents an acre in the 1950s when Jacobs started. Unlike a mining claim on which there is no fixed rental but which requires at least $100 worth of development to be carried out within a year of staking it out, there is no necessity to do any work on an oil lease. It will lapse after three years, unless renewed. The usual area of a lease is 2,560 acres, granting the holder ownership of any petroleum that might be found on it. In practice, unless an individual is wealthy enough to carry out exploratory drilling himself, he will hope to sell the lease to an oil company for a profit—at anything between one to a hundred dollars an acre, depending on the prospects—and retaining an overriding royalty of perhaps 2 per cent on any oil found and produced. An oil company does not have to hold leases to carry out preliminary survey work in a particular area, but naturally if there is a chance of oil being present it would want to lease as much of the surrounding land as possible. If this is already held by individual speculators, then their profits can be limitless. But plenty of people have lost their money by leasing unproductive land, or by not being able to keep up payments until oil is found. It is truly a gamble, and in fact may not last much longer for the federal government is now considering a whole new leasing structure.

The only exceptions to federal lands that may be leased in this way are those reserved for specific purposes, such as naval petroleum reserves, wildlife ranges or Indian reservations, or those lands which are 'known geologic structures' according to the US Geological Survey, where previous investigation and maybe drilling has indicated the certain presence of oil-bearing formations—although that does not mean there must necessarily be commercial quantities of petroleum. In the latter case, and it also applies to state lands, leasing is done either by competitive bidding or simultaneous filing, two systems which are explained in a later chapter. As the past history of the United States shows, even reserved areas of the former kind are not inviolate to the entry of the oil industry if the political pressures are great enough, but they are not normally available to leasing by individuals.

At one time three-quarters of the United States' total land mass of over two billion acres was under federal ownership, including the whole western area. But as the various states came into being they gradually took over much of it as state land, with the right to charge taxes and control its exploitation as they wished. Naturally, such plum areas as known petroleum prospects were among the first selected, on which the state could impose and collect its own royalties on any oil produced; in the event of it still being federal land, the state received only 90 per cent of what the federal government collected under its own tax laws. It was not to the advantage of a state to take over lands until they were needed or could produce an income, however, for in doing so it would lose federal grants for such public services as fire-fighting and highway construction, the costs of which then had to be paid by the state. In 1958, the year before statehood, Alaska's 375 million acres accounted for nearly half of the land in the United States that was still federally owned—some 770 million acres. Even now, partly because of the 'land freeze' that has been imposed in Alaska until the native claims are settled, more than 93 per cent of Alaska is still federal land.

Alaska might have been an oil speculator's paradise—except that in fifty years of intermittent exploration and drilling no worthwhile quantities of oil had been discovered. Interest had fallen off so sharply that in 1949 there was only one lease of 2,560 acres in force on public lands, as against fifty-nine leases the year before. As a result of the sporadic activities by oil companies in the Kenai and Alaska peninsulas which had so intrigued Atwood and his friends, further leases were taken out over the following years, but they had still only reached 139 by 1952. As chance would have it, one of these was at Iniskin Bay, the exact location on which Locke Jacobs filed his first lease. It took the Bureau of Land Management six months to find this out and let Jacobs know it was already covered, at which point he realised that no one had any proper maps of oil-leases in Alaska. So he decided to provide a leasing service himself, laboriously copying the records by pencil—later by typewriter when he could afford it. All this was in his free time, for he was still working at the surplus store. It wasn't long before his boss got to hear about it—especially as Jacobs was eagerly promoting the idea of lease filing amongst the other employees. He strongly advised him against such a risky venture as oil leasing.

'He asked me what I knew about oil,' Jacobs says. 'I told him, not very much—but then neither did anyone else in the state. I guess I convinced him—or maybe he just wanted to pay me to stop talking about it—because he wrote me out a cheque for $1,000 to get some leases for him.'

Jacobs filed on several leases and soon afterwards sold them to Shell Oil, doubling his boss's money. For part of the game is that the oil companies must keep options over millions of acres of available land, even if they are not likely areas, just in case anything is discovered.

They juggle with these as prospects brighten and dim, for oil has an infuriating habit of not turning up where it should and then being found where least expected. The Shell landman was so impressed by Jacobs' drive and enthusiasm that he hired him as an oil scout. Jacobs' boss was also so impressed that he told his friends at the 'Spit and Argue' club. And that was the turning point for Jacobs, for the businessmen who wanted to promote economic growth in the area, for the oil industry, and indeed for the whole state of Alaska. Jacobs was just the man that Bob Atwood and his colleagues needed, someone who knew about leasing and who could help to handle their project for them. Between them they filed leases on over 300,000 acres of the Kenai Peninsula around the Swanson River area.

The intention of Atwood and his group was not primarily to make a profit but to try to get an oil company interested in drilling a well to test the possibility of oil being present. If there was, it would mean a new boost to the economy and lay an important foundation for statehood. It was on the advice of a friendly geologist that they plumped for the Kenai Peninsula, which he described as a good oil prospect. But he was the only one who did, as Jacobs found when he did the rounds of the oil companies in the lower forty-eight.

'I got kicked out of a lot of companies when I was trying to peddle the leases down there,' he says. 'They'd tell me—"What are you trying to do, sell us a glacier?" They thought of Alaska as just a place of ice and Eskimos.'

Even those companies whose field parties sent to Kenai over the years had reported favourably on the area were not interested. To give them their due, it would be a costly decision to drill a well, especially in Alaska which had proved to be barren for so many years and where none of the usual oil-well services existed. Everything would have to be brought in from outside. It was one thing to file leases or send geological parties into an area—it was quite something else to embark on a drilling programme. Companies can over-analyse an area to such an extent that they talk themselves out of thinking it might be a good prospect. That seems to have been the case with Kenai, for the only company that showed any interest was Richfield Oil—the one company that had never done any field survey work at all there. Richfield at that time was a small company, a poor relation of the industry. Desperate to keep the company's interest when they appeared to be having second thoughts, the Anchorage group offered to let them have an option on the leases for nothing.

'They just couldn't understand it,' Atwood recalls. 'No one had ever offered them something for nothing before. In the end they insisted they had to buy them, so we agreed a price that was twice what we had paid, including a penalty clause that if they drilled a well, it would be on our leases.'

Richfield carried out a survey and decided to drill a well. But the site they chose as seeming to be the best location was not on the group's leases but on one of their own, nearby. The company wanted to know how much the group wanted to waive the penalty clause. The businessmen got together, in the powerful position of being able to hold Richfield to ransom over the clause. But they decided the most important thing was to get the well drilled. They agreed not to press the penalty clause. All seemed set—until it was suddenly discovered that Richfield's lease was on a moose range.

This had been set aside as a protected area for a herd of some 4,000, for although moose are fairly widely distributed over Alaska, totalling about 130,000, they are also one of the most hunted animals and the easiest to stalk and kill. There was nothing against anyone filing leases on such a range; it simply meant that the Fish and Wildlife Service imposed restrictions on how any industrial development should be carried out, with the least inconvenience to the animals. However, it was just at this time in the mid-1950s that controversy had arisen about leases filed on a wildlife range in Louisiana. Filing was stopped on all such reserves while Congressional hearings were held to review the exact situation on each one. The oil industry as a whole in the United States was coming in for a lot of criticism at that time, and Richfield didn't want to get politically involved. It looked as if the well was not going to get drilled after all.

So Atwood and his friends decided to 'do it the Alaskan way' by taking the matter up for themselves. They got sportsmen and conservation societies to agree that drilling would not hurt the moose. Armed with this, the group tried to get a permit to drill from the Interior Secretary, who was then Fred Seaton in the Eisenhower administration. But no luck there. The Congressional hearing had established that drilling would not harm the moose, but an even bigger issue had developed over the trumpeter swan, a bird close to extinction —no more than 200 existed in the whole of the United States—which migrated each year to nest on the lakes in the reserve. So the group had another idea. They had found that if anyone could get what they wanted, it was usually the military. A long report was prepared on the security need for the military to have locally available oil, rather than having to bring it in by pipeline. In this way it was hoped to get the Army's backing to drill the well. Atwood and a colleague went to Washington to see General Twining, the Chief of Staff, who they both knew since he had at one time been Commander-in-Chief in Alaska.

'Our idea was to see Twining first and give him the report.' Atwood again. 'Then we'd see Seaton who we hoped would give his approval. But the way the appointments worked out, it was Seaton we saw first. We were a bit worried, but after the situation had been explained, Seaton said it was okay if Twining agreed. He'd give him permission to drill on the White House lawn if he wanted. Then we saw Twining in a room filled with maps and trophies of Alaska—he loved it when he was up here. There was a whole bunch of colonels in there, and it ended with Twining giving them reasons for needing oil on Kenai.'

The permit came through, although exploration was not allowed on those lakes where the trumpeter swan nested. In order to move in a drilling-rig it was first necessary to build a twenty-five-mile road into the reserve. Atwood recalls with amusement that this was done by unwinding a roll of toilet paper from an aircraft flying low along the ridges, which a bulldozer then followed by tracking the line of paper as it came to rest in the tree-tops.

By this time, other companies had woken up to what was going on in the Kenai Peninsula. Standard of California, Atlantic Refining, Union, Marathon and Phillips, among others, all began to lease land in the area as a precaution, either through Jacobs' leasing service or taking options from him on land he had already leased. He was now working full-time on this, having given up his clerk's job, and was reinvesting any money he got in further leases. By the beginning of 1957 more than 3,000 leases were held in Alaska, not only in the south-central area but in other parts as well. And then on July 23 Richfield's wildcat well, fifty miles south-west of Anchorage on the banks of the Swanson River, struck a large deposit of oil at a depth of 11,131 feet. It was Alaska's first major discovery and set off an oil boom not unlike that which had followed Drake's original strike nearly a hundred years before, although the interest in it was not yet on the international scale which came a few years later with the North Slope finds. The companies were queueing up outside Jacobs' office. During the weeks that followed he was making $1,000 an hour for as long as he could keep awake, employing eighteen secretaries and four draughtsmen just to make maps and charging ten cents an acre to make out leases. Within a day of the announcement of the discovery he had filed on about one million acres of land, cornering the market in many areas. By the end of the year, the number of leases held in Alaska had shot up to over 9,000, covering nearly twenty million acres.

Important though the Swanson River discovery was, it had not yet proved the existence of a major oilfield. That would require further test drilling and at a million dollars a well Richfield could not afford the development costs by itself. As it later transpired, the structure was an elongated shape—had Richfield drilled its first well a quarter of a mile away on either side, it would have been dry, in which case it might have been years before another attempt was made. Although they didn't know it at the time, Atwood and his group owned the leases over most of the actual oil reservoir. These they sold for a good profit to Standard of California. Then Standard and Richfield came to an agreement for the joint exploration and development of their holdings with Standard as the operator; and when the group heard how much it had cost Standard to buy into the project—some thirty million dollars—they wondered if they hadn't been taken for a ride. However, they had achieved their main goal of bringing the oil industry into Alaska. It was a major stimulus to the granting of statehood two years later.

It wasn't until 1960 that the Swanson River field came into production at an initial 20,000 barrels a day, rapidly rising to 30,000 barrels a day. It had taken four development wells to find the main part of the structure and there were gloomy forebodings that the first discovery might have been a freak. Drilling expenses were made higher by the extra work required to protect the wildlife environment—in fact, as a result of revegetating vacated land with grass seed, the number of moose on the reserve actually increased. One problem arose when the federal government tried to impose a penalty of $25,000 on any moose run over and killed by an oil company vehicle. The companies protested but would have gone along with it, wanting to avoid trouble. But again the Anchorage businessmen came to their aid. They felt the penalty was grossly unfair and persuaded the governor of the state to say that if it was enforced, then a similar penalty should apply to the railroad—and in one year trains had killed no less than 250 moose. That was the last ever heard of that suggestion.

One of the big advantages that attracted oil companies to southern Alaska, once oil had been discovered, was its closeness to the oil-hungry markets of the US West Coast, especially California, which did not have sufficient production of its own to meet demand. Also, the relatively ice-free waters made it possible to ship the oil by tanker all the year round. Exploration progressed rapidly in the search for more oilfields like Swanson River, which was established as a 5,000-acre structure with about 175 million barrels of reserves. A twenty-two-mile pipeline linked the field with a marine terminal built at Nikisiki on the Cook Inlet, from where shipments were made to the West Coast refineries. A small refinery was built near Anchorage to meet Alaska's oil needs. But

ABOVE The spectacular Portage Glacier, fifty miles south of Anchorage. Floating icebergs in the 600-foot-deep Portage Lake often weigh several million tons.

BELOW Anchorage, Alaska's largest city, with a population of about 130,000. The open area in the foreground is where whole blocks of office buildings collapsed in the 1964 earthquake disaster.

ABOVE Surrounded by the rugged mountains of southeastern Alaska, Juneau, the state capital, is inaccessible by road and can only be visited by plane or boat. With one of the highest rainfalls in North America, protective coverings are built over most of the sidewalks.

BELOW Summer at Sag Valley Lake in northern Alaska, when the sun never sets. This photograph was taken at one o'clock in the morning.

ABOVE Lone prospectors, such as Charlie Bilderbach, still pan for gold in fast-flowing streams.

BELOW The Brooks Range, rising from the North Slope. It is across these mountains, many of them still unexplored, that the trans-Alaska crude-oil pipeline must pass.

COLT

OPPOSITE, TOP Don Sheldon, one of the last of Alaska's bush pilots, refuelling at Talkeetna.

OPPOSITE, CENTER The old way of transport, by dog sled . . .

OPPOSITE, BOTTOM . . . and the new, by snowmobile.

THIS PAGE, TOP A geologist collects rock samples at the edge of the Malaspina Glacier, the largest in North America.

THIS PAGE, ABOVE To the beat of drums and the rattle of tambourines, Eskimos perform one of their traditional dances for tourists visiting Barrow.

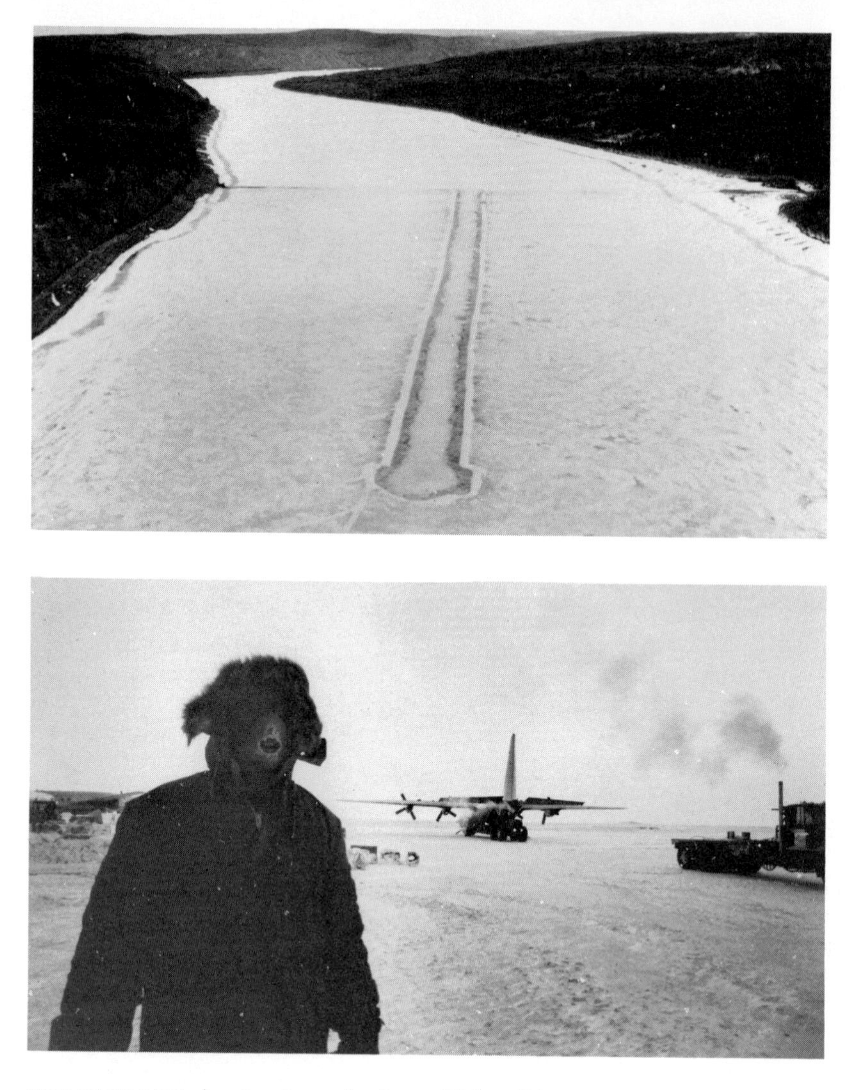

THIS PAGE, TOP An airstrip on the frozen Yukon River, near the crossing point of the trans-Alaska pipeline.

THIS PAGE, ABOVE Monster of the Arctic: wearing a face-mask for protection against frostbite, an oil-man helps unload a Hercules aircraft bringing supplies to a North Slope drilling site.

OPPOSITE, TOP Legacy of the oil industry's invasion: fuel drums strewn about a camp-site on the North Slope.

OPPOSITE, CENTER Hundreds of miles of pipe remain stacked, waiting for the final decision on building the trans-Alaska pipeline.

OPPOSITE, BOTTOM Part of the winter trail, bulldozed through valleys and over hills from Livengood, just north of Fairbanks, to the North Slope. This will be the route of the trans-Alaska pipeline.

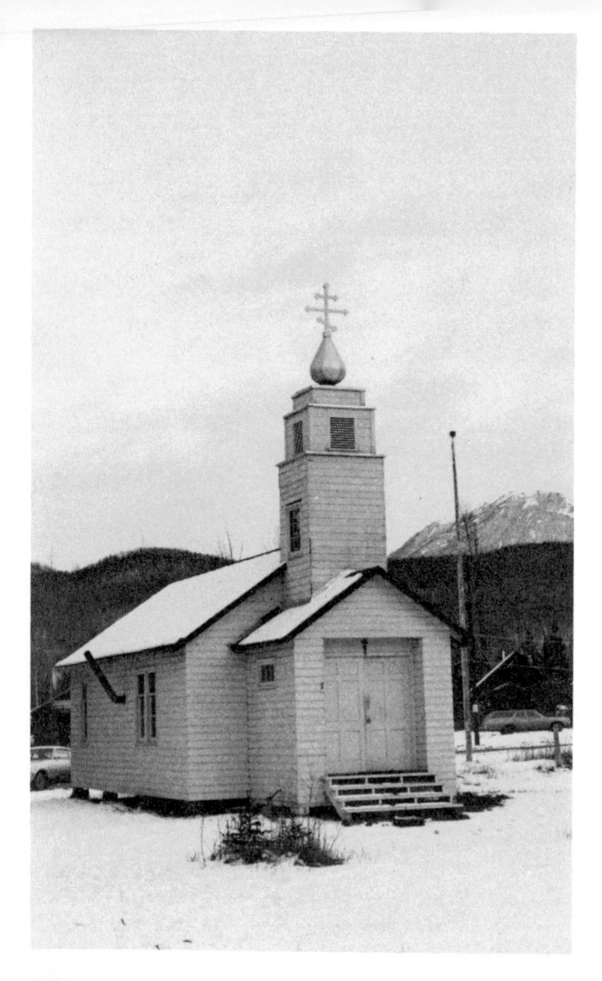

LEFT A tiny Russian Orthodox church at an Indian settlement in southern Alaska, with traditional onion-shaped spire. Orthodox Christianity among the natives of Alaska is the main legacy of Russian occupation.

BELOW Even older, however, is the charming Indian custom of building gaily painted miniature huts over graves, in which are placed various implements and food to help the deceased in the next world.

Alaskan wildlife: A young bull moose (*above*) grazing beside the Echooka River and a Toklat grizzly (*below*) at kill in McKinley National Park.

ABOVE Dall sheep on the mountain slopes of Atigun Canyon.

BELOW A grey wolf on the North Slope tundra in summer.

four years went by before another oilfield was discovered. Most of the early wildcat wells found natural gas, some of it in large quantities, but there was no immediate market for it except locally in Anchorage. Gas from the Kenai gasfield, for instance, discovered by Union and Marathon in 1959 some twenty miles south of Swanson River, was piped to Anchorage, which two years later changed over from a fuel-oil to a natural-gas economy.

In the early 1960s the search mushroomed as new companies moved into the area to join the exploration already being carried out by Standard of California, Union, Marathon and Halbouty. The sedimentary basin was known to extend under the whole of the upper Cook Inlet, from Kenai Peninsula to Alaska Peninsula on the other side, and it was not long before the decision was made to leave the shores and concentrate on offshore exploration under the waters of the Cook Inlet itself.

From the map, Cook Inlet appears to be a well-sheltered waterway, with the name of Anchorage implying a cosy haven of safety. What the companies found instead were the toughest offshore drilling conditions they had ever encountered. The waters are thick with glacial silt, making it literally impossible for a diver to see his hand in front of his face so that all underwater work must be done by a sense of touch. Although water depths are fairly shallow, not more than 125 feet, tides rise and fall by as much as thirty feet. Tidal currents sweep in and out of the Inlet at eight knots or more, carrying with them in winter large ice-floes which can crush hard against moored vessels. Ice-fogs and blizzards also add to the hazards in winter. Nevertheless, in 1962 the first drilling platform was built in the Inlet, a huge man-made island of steel rising up from the sea-bed to the height of a thirty-storey building above the water. From this and other platforms which soon followed, wells could be drilled and angled as much as 6,000 feet from the rig itself, so that there was no need to move a platform in order to drill another location. In that first year of offshore operations three wells were also drilled from floating vessels. The result by the end of the year was that one well had blown out but was later successfully abandoned, two were suspended because of the unfavourable climatic conditions, and the fourth was still blowing wildly, out of control. There was no doubt as to the presence of natural gas, nor to the difficulties of operating in the Inlet.

In 1963 a Shell-operated group including Richfield and Standard of California struck oil in the Inlet for the first time, confirmed as an oilfield the following year by a Pan American discovery in the same structure and named the Middle Ground Shoal. Because of the high costs involved, most of the companies had banded together in various groups, one company being nominated to undertake operations on behalf of the others. The most successful discovery period of all came in 1965, when within a few months no less than three oilfields were located. The biggest was the McArthur River field, discovered by Union Oil of California as operator for itself and Marathon. Mobil Oil, operating for itself and Union, discovered the Granite Point field, so named because it lay three miles from a great bluff of glacial boulders on the western shoreline of the Inlet. And Union again, as operator for itself, Marathon, Texaco and Superior Oil, discovered the Trading Bay field. By the beginning of 1968 these companies, including a participation by several others, had in operation eleven fixed drilling platforms in the Cook Inlet, most of them bearing twin rigs. Fifty production wells had been drilled to tap the four oilfields, averaging more than 2,000 barrels a day each, while the forty wells at Swanson River were averaging about 900 barrels a day each. When total production peaked at over 220,000 barrels a day later in the year Alaska was ranking eighth amongst the oil-producing states of America. The Cook Inlet basin was reckoned to contain 1·5 billion barrels of oil reserves and five trillion cubic feet of gas. All this was heady stuff to the new young state of Alaska. Oil was bringing in $50,000 a day in royalties and taxes, leases in the Cook Inlet were selling for as much as $1,000 an acre. From 1957 to 1968 the state collected $140 million from the oil industry. Petroleum had become Alaska's third most important industry after fishing and timber.

But there were troubled waters which oil could not soothe—just the reverse, in fact. Traditionally jealous of their state's natural beauty, Alaskan conservationists were becoming increasingly outspoken in their criticism of the oil industry. They had accepted the idea of drilling on the Kenai moose range, where Standard of California had set a very high standard of environmental housekeeping. The wildlife had not been harmed and the incursion of the oil industry had in fact opened up large areas for hikers and canoeists which would not otherwise have been accessible. But in the waters of the Cook Inlet the problem was much greater. However carefully controlled, it was inevitable that seismic explosions would kill fish. Accidental spillages led to water pollution and beach contamination, to the point where by 1968 over sixty instances of oil or refuse pollution had been reported, leading to an official statement that 'they have served to tarnish the industry's Alaskan image'.

From the oil industry's point of view, the picture was not all that rosy either. The early success in discovering oilfields had not been maintained; meanwhile, the Cook Inlet fields had not proved to be as productive as was initially thought. Pressures were not great enough to increase production except by the drilling of additional wells, and in some cases, outputs were beginning to fall. The companies were operating very much in the red—investment by 1968 totalling nearly $500 million as against a return of $250 million. A break-even point is not expected to be reached until the mid-1970s. Impressive though it might have seemed that Alaska ranked eighth amongst the oil-producing states—and it certainly was an achievement in only eleven years—the figure of 220,000 barrels a day dwindles in comparison with the highest of over three billion barrels a day from Texas.

To those oil companies operating primarily in the United States, accustomed to small oilfields and high crude oil prices that make them worth developing, together with tax advantages that considerably reduce the cost of exploration and production, the Cook Inlet was nevertheless a reasonably attractive area of operations. From the point of view of a major international oil company, however, used to much larger oilfields in the Middle East and elsewhere—Kuwait's Burgan field, for instance, alone produces 1·8 million barrels a day—the prospect did not look very promising. Such a company was British Petroleum, the only one of the majors without a stake in the United States and desperately keen to get one. BP looked at the Cook Inlet play and did in fact take a share in several exploratory wells drilled there up to 1963, in addition to partnering an unsuccessful 12,000-foot test well near Nulato in the lower Yukon basin. But the company was not attracted to the area, partly for geological reasons but mainly because so much of the land was already leased. It was not accustomed to the wheeling and dealing necessary to obtain options from a multitude of individual leaseholders—the 'shoe clerks' of Anchorage, as BP's chief geologist wrote rather acidly—as against the relatively simpler method in other countries of leasing direct from governments.

Locke Jacobs smiles indulgently at the phrase 'shoe clerk'. He can afford to—all the way to the bank. From his luxurious Anchorage office overlooking Cook Inlet, complete with polar bear skin on the floor—'I didn't hunt it, guess he just came in and got tired'—he has so far in his meteoric career covered over thirty million acres in leases. He has holdings in oil-wells and in real estate in many parts of the United States. His shelves of files and maps record every lease in Alaska and its ownership. He is now turning his attention to mining and prospecting for other minerals. It is likely that as with petroleum, he will be first there when the expected mining boom occurs in Alaska within the next few years. For him, the American dream came true.

BP meanwhile, too late to buy directly into the Cook Inlet and unwilling as a newcomer to get involved in the rough and tumble of lease-dealing, turned its attention to the northern part of Alaska which had largely been ignored by everyone else. In doing so it laid the foundation for its own dream of a stake in America.

10

The British Invasion

The elderly couple from Fairbanks had flown up to the North Slope in their small Cessna aircraft and landed on the coast near the mouth of the Colville River. There they pitched camp and spent several days studying the birds, who for the brief two months of the Arctic summer were making the wilderness their home. It was mid-August 1963. The temperature was a few degrees above freezing. Against the pale blue sky and the slanting sun, which never sank below the horizon at that time of the year, the air was alive with the ballet of gulls, kittiwakes, sandpipers, plovers and many varieties of ducks and geese, their strutting and whistling courtship rituals making up for a less than exotic plumage, mostly protective shades of black, grey and white. There was the Arctic tern, the world's longest migrator, which nests on the tundra and winters in Antarctica; the American golden plover which migrates from the North Slope to the south coast of South America. The tundra itself had shed for this brief period its mantle of ice and snow. Between the pebble-pattern of myriad ponds and lakes which made such ideal nesting sites for watefrowl were carpets of moss, lichens, poppies, fireweed and showy-flowered dwarf shrubs, struggling to make the most of their moment in the sun. Offshore the ocean ice had broken up and the sea was grey-blue between the floating remnants of ice-floes. A strange and beautiful land, desert-flat as far as the eye could see in any direction, empty but for the two birdwatchers and the birds themselves and the tiny lemmings burrowing in the vegetation and, way off towards the mountains of the Brooks Range, herds of browsing caribou following their ageless instinct to migrate northwards every summer.

Or was it empty? There came an afternoon when the couple returned to camp and saw to their astonishment a tug turning from the ocean into the Colville River, several long barges strung out behind it. Nothing like it had ever been seen on the North Slope before, and their first fearful thought was that the Russians were making an invasion with landing craft. It was not perhaps such a surprising reaction, considering that in this most northern part of the United States were located the advance defences of the DEW-line early-warning system, set up against the possibility of just such an invasion. Had the couple followed their first instinct, and sent out an urgent warning over their radio, there is no knowing what the repercussions might have been. But they hesitated long enough to become aware of the crew waving to them from the tug and to make out the Canadian flag drooping from the mast. It was not a Russian invasion, but it was an invasion of a kind. The barges were bringing in a drilling rig and equipment for British Petroleum, who were planning to search for oil on the North Slope. It was the first time that a British oil company had ever attempted a major oil exploration in the United States. If it were not for BP there might not have been a North Slope oil boom. The British venture had its origins twelve years before on the other side of the world, in Iran.

As one of the seven major international oil companies, BP's strength had always been in the Middle East. Its vast reserves of oil in Iran, Kuwait and Iraq were greater than those of any other single company, over a fifth of the total reserves of the whole world in fact. This was a strength but also a weakness, for while the other big companies were developing their marketing outlets throughout the world BP was content to rely mainly on its profitable oil production, selling any crude oil that it could not market itself to other companies who had the markets but not the production. As far back as the mid-1930s BP had thought about entering the United States, the world's most lucrative oil market and one in which all the other major companies were operating, including Royal Dutch Shell through its associated company Shell of America. But there always seemed to be a reason for putting it off. And then after the Second World War, when BP decided it really did want to go into America, the big problem was Britain's adverse balance of payments and the corresponding shortage of dollars. It would have been an immensely costly investment either to buy up existing companies or to start from scratch with all the production and refining and marketing operations required. And in any case BP was doing very well as the biggest producer of Middle East crude oil and possessing a lion's share of its reserves.

This cosy security was shattered in 1951, however, when, under the xenophobic leadership of the Iranian prime minister, Dr Mussadiq, Iran suddenly nationalised her entire oil industry and BP was compelled to withdraw after fifty years of first entering the country. It was the first major act in the tide of nationalism that was beginning to sweep across nations previously dominated by foreign industries. In the event it did not do Iran very much good. The Iranians were unable technically to run such a complex industry, and during the legal battles that followed an embargo was placed on anyone accepting her oil. In 1954 BP returned to operate in Iran as the major partner in a consorttium of American and European oil companies. But it had been a traumatic shock, jogging the company's management out of its rather complacent attitude towards marketing and proving the need to develop oil production in other parts of the world. In that sense it was the best thing that had ever happened to BP, resulting in determined expansion in place of a tendency towards self-satisfied stagnation—'kicking us smartly into the twentieth century' in the words of one BP executive.

Within weeks of being forced out of Iran, a group of BP geologists, led by Alwyne Thomas, C. A. O'Brien, Frank Slinger and Harry Warman, got together to prepare a world survey of oil prospects based on their knowledge of the sedimentary basins in various regions. These were graded according to the likelihood of oil being present. The survey was completed early in 1952 for consideration by Peter Cox, then head of BP's exploration department, and eventually led to major oil discoveries being made in Nigeria, North Africa, and offshore in such areas as the Persian Gulf and the North Sea. Amongst the several hundred places given a number one grading was the North Slope of Alaska, primarily because of the discovery of the Umiat oilfield which had just been announced by the US Navy. But the area was too remote to be given serious attention, and in any case there were more worthwhile opportunities elsewhere. Alaska was for the time being forgotten. The closest the company got to that area was in 1953 when BP took over the Canadian Triad oil company operating in Alberta.

It was another political leader in the Middle East, the late President Nasser of Egypt, who provided the final impetus for BP's entry into the United States. The Suez Canal crisis of 1956 convinced the board of the company's weakness as compared with those who had a large share of the profitable and stable North American oil industry. By supplying Europe with North American oil when supplies from the Middle East were cut off, these companies benefited from precisely the kind of circumstances that were most damaging to BP. But the cost of going into the United States directly was higher than ever, and as far as

Latin America was concerned there were prohibitions against granting concessions to any company in which a foreign government held a majority shareholding—and the British government at that time owned 56 per cent of BP's shares (although this was later reduced to under 50 per cent by successive share issues which the government elected not to take up). The answer to the cost problem seemed to be in finding a company which was already marketing in the United States but which was short of crude oil, and coming to a joint agreement whereby BP would supply it with oil from its abundant Middle East supplies. Consultations were held with a number of US companies; meanwhile BP's operations in Canada were expanded and in 1957 the company took over Rio Tinto's interests in the Western Hemisphere in return for BP shares. What attracted BP mostly were several small producing oilfields in Trinidad and Canada which it was hoped could be expanded. There were also some producing wells in the Kern River field area of California. Although no one knew much about these at the time, they were to become a vital factor in the Alaskan operation later on.

The small office in New York's Rockefeller Plaza which BP ran as a kind of advance post in those days saw a constant succession of meetings between British and American oilmen to discuss various deals. Eventually, in 1958, BP came to an agreement with a medium-sized American company, the Sinclair Oil Corporation, which marketed products mainly in the eastern states but which was deficient in crude oil sources of its own, having had a marked lack of success in its exploration efforts. BP was to fill the gap by importing supplies from its Middle East oilfields, following the trend already set by those American companies who had discovered oil in the Middle East and Venezuela and were bringing it into the United States in increasing quantities. The prize was the highest-priced oil market in the world. Through prorationing regulations, domestic production was limited to actual demand in order to conserve supplies and not to over-produce as had been the case in the early days of the industry. But this also had the effect of keeping price levels high and encouraging the continued operation of wells that, by world standards, were inefficient and uneconomic. The cost of producing oil from the many thousands of marginal stripper wells in the United States that pump out no more than an average of about five barrels a day—and these account for two-thirds of America's total 573,000 producing wells in fact—was over a dollar a barrel, compared with about ten cents a barrel in the prolific Middle East oilfields. Whereas domestic oil sold on the East Coast for around $3·40 a barrel, the landed price of imported oil was no more than $2·00 a barrel, including ten cents import duty. The American consumer would benefit from these lower prices of course, but at the same time the United States would become increasingly dependent on overseas oil supplies which might become cut off in a political crisis, as European countries had found to their cost.

The BP-Sinclair agreement was frustrated from the very outset. Scarcely had the contracts been signed when the US government, alarmed at the threat of low-cost foreign oil flooding the market, imposed at the beginning of 1959 rigid controls on crude oil imports which were not to exceed 12·2 per cent of total domestic production. Only those companies with a domestic production could import foreign oil, effectively excluding BP. In an attempt to be fair the quotas to buy imported oil were distributed to refiners on the basis of a percentage of their throughputs. This effectively gave some companies who were not even geared to use imported oil—and were too far away from the coast to have taken advantage of it anyway—import tickets worth between $1.25 and $1.40 a barrel which they simply sold for a cash sum to refiners who could make use of foreign oil. It was a windfall at the expense of the American consumer. In spite of bitter criticism by the consuming states against the producing states who gained through prices being kept at a high level, the import programme still operates. But it has run into many difficulties. So many exemptions have been allowed, such as unlimited imports of residual fuel oil, special allowances for petrochemical manufacturers and others, that total crude oil and product imports now supply about 23 per cent of US demand and will likely reach 30 per cent by 1980. With domestic production unable to keep up with a fast increasing consumption the United States no longer has a choice but must of necessity import greater quantities of oil.

Unable to market Middle Eastern oil in the United States, BP and Sinclair turned to the other part of their joint operation which was to explore for new sources of oil. They had already discovered a small oilfield in Colombia—BP could now operate in Latin America since it was in partnership with Sinclair. By this time Richfield had made its Swanson River discovery in Alaska and the US Navy had published the results of its oil exploration on the North Slope. Sinclair suggested the possibility of looking for oil in Alaska. What it had in mind was the southern Cook Inlet area where many American companies were now operating. But BP, recalling its own survey of world oil prospects made in 1952, was more interested in the North Slope. This reflected a fundamental difference in approach between BP's geologists and those of Sinclair and other US companies. Their thinking was geared to finding small oilfields which, because of the high price of domestic oil, were profitable to develop. BP, on the other hand, was accustomed to the much bigger scale of Middle East exploration where the oilfields were many times larger. One of BP's executives in New York in 1958 was Robert Belgrave, who remembers it this way:

'Peter Cox was our chief geologist then. He came into my office one day and said that the North Slope was one of the few places in the world where there might be a chance of finding a Kuwait-size oilfield.'

Cox had just come back from a visit to the North Slope, and had been impressed while flying over the Brooks Range at the similarity between the foothills there and those of the Zagreb mountains in Iran where the first Middle East oil had been found, leading to the birth of the BP company. The North Slope got its name, in fact, from the way the foothills sloped gently towards the Arctic Ocean. It was similar in many geological respects to the deserts of the Middle East—except for the difference of about 150 degrees of temperature—and the annual four inches precipitation on the North Slope was even less than in most deserts. Cox shortly afterwards became BP's managing director of exploration and it was he who, in February 1959, accepted the recommendation of a report to London by BP geologists which stated: 'The areas of the sedimentary basins are very large. That of the Arctic Slope measures 105,000 square miles, larger than our entire Iranian concession. . . . It contains a wealth of drillable anticlines on the Iranian scale, with lengths of the order of twenty miles. . . . Overall prospects are of a high order. . . .'

So BP decided for the first time to embark on exploration in the United States. Because of anti-trust laws this could not be done through the joint BP-Sinclair company, although the two companies could bid together for acreage and share technical data. Sinclair already held some leases on the North Slope, as did a number of American companies, taken out as a protective measure when the federal government first opened the area for leasing in 1958. Shell of America owned a large amount of acreage there, but allowed it to lapse. Like all the other American companies it was more interested in exploring the Cook Inlet, especially after Swanson River was proved a productive field in 1959. Since at that time an individual company was limited to a total holding of only 300,000 acres in the whole of Alaska, it meant giving up northern leases in order to take part in the scramble for land in the south. And in any event none of the American companies could see any future for the North Slope. As Frank Rickwood, then one of BP's geologists and now head of BP Alaska in New York recalls: 'I was always embarrassed when American companies said to me, "What in God's name are you doing in that wilderness?" I remember when the

North Slope first came up I had to send a girl down to find a map of Alaska to see where the bloody place was.'

The fact that no one else had any interest in the North Slope was a great help to BP, enabling the company to pick up leases cheaply in the foothills of the Brooks Range near the US Navy's discoveries at Umiat and Gubik which then seemed to be the most promising locations. But it was vital that this should be done secretly, since any movement by a major oil company would attract the attention of other companies and promoters, leading to sharply increased competition for leases and higher prices. BP was a novice in the complicated American game of oil-leasing. At the outset one of Sinclair's contributions had been to provide detailed maps of land holdings which showed who held exploration rights and what acreage was still free, a service which was a great help and which greatly impressed BP until they found that such maps could be purchased by anyone for twenty dollars each. A retired American oilman, formerly a vice-president of Pan American Oil, was employed to apply for leases in the name of BP at the land office in Fairbanks, the seventy separate forms all in quintuplicate being typed confidentially in a local bank. These were filed on April 1, 1969, and BP was firmly established on the North Slope with 50,000 acres of leases and options on another 100,000 acres. It was time to send in the geologists who would be the vanguard of the exploration effort.

One of the geologists closely involved from the start was Geoff Larminie, a shrewd, impish Dubliner who has been to many strange corners of the world in the search for oil and has since become BP's area manager in Alaska.

'We started work in 1959 in the Richardson Mountains and went to the North Slope the following year. By 1966 we had covered it back to the Brooks Range. The geological work had to be done in summer because you've got to let the snow melt so that you can see the rocks. We gradually worked northwards as the snow melted, setting up camp alongside lakes so that our float planes could land. The weather was terrible sometimes. There were days when we were stuck in our tents, unable to fly, and had to keep warm by staying in our sleeping bags all day. Because of the weather, flying in the mountains was very tricky. Fog and low cloud could suddenly seal off a whole valley. Everyone had to carry emergency rations and the orders were to stay put if you were cut off until you could be rescued. Shortage of water was a problem, strangely enough. People could become badly dehydrated if left out in the mountains with no rations. You could be stuck out there with your head in the sky and your backside in the snow for days on end. Usually the camp would wake up at about six o'clock. Someone would make breakfast, then the first helicopter would be away with the first team—we normally worked in parties of two. It would come back for the second team, and so on, until we were all out scouting the hills and valleys.'

In that way the patient work went on day after day of collecting and examining the rock samples and making careful notes in weathered field-books. It was not long after BP started that other companies entered the area to conduct similar operations, attracted both by BP's obvious interest and the US Navy's exploration reports. Sinclair was there already, of course, in a joint venture with BP; then there were such companies as Humble, Atlantic Refining and Richfield, together with teams from the US Geological Survey who were systematically mapping and studying the whole area. Parties of geologists went all over the northland, examining outcrops and chipping rocks to send back to laboratories where the fossils in them could be identified, gradually building up a geological picture of the region. It was dangerous work; every year at least one geologist was killed during the season, either in helicopter accidents or by falling over cliffs. One USGS geologist walked away from three helicopter crashes in one season and vowed he would never come back again. Another geologist from the same organisation, George Gryc, who had also worked before in the Naval Petroleum Reserve, tells the story how one year they decided to bring ponies up from Montana and travel by canoe and pony instead of by plane. At the end of the season they found they were minus one pony, but they had to move out and there was no time to look for the animal. Several years later Gryc was up on the coast talking to an old Eskimo he had known for some while and who had often helped the geologists. The Eskimo was very excited at something that had happened during the winter. He had looked out of his shack during a snow storm and seen a huge monster rearing up and making a terrible noise.

'I got my gun and shot it,' the Eskimo said. 'It was not good to eat, but the strange thing . . . Here, let me show you.' He took the geologist to the back of his shack. 'The monster had feet of iron,' the Eskimo whispered. 'Never before have we seen anything like this.' There on the wall were four horseshoes, all that remained of the lost pony.

While the preliminary survey work was going on in the north, BP was also involved in operations elsewhere in Alaska. An area along the southern coast near Yakutat was prospected in partnership with Colorado Oil and Gas. Two wells were drilled further to the north-west near Katalla, initially with Sinclair and later by BP alone. The company also had a share in a well drilled in the lower Yukon River basin at Nulato by a group headed by the Benedum organisation of

Pittsburgh. But these were of secondary concern. The North Slope was the main objective where BP and Sinclair had obtained an excellent spread of leases covering the most critical areas in the foothills north of the Brooks Range. It was a desolate region, far north of the treeline, the rocks on the hillsides broken down by the perennial frost into a dismal grey scree. Air photographs and geological and seismic surveys had by 1963 located promising structures ten to twenty miles long with such descriptive names as Shale Wall and Outpost Mountain, and Eskimo names like Kuparuk and Itkillik. Some of them had their reservoir sandstone beds exposed so that the centre of interest lay in the covered folds further north, near to where the US Navy exploration had found a gasfield at Gubik and a small oilfield at Umiat.

All this had cost money, of course, and the early 1960s were lean years for BP, both in terms of cash flow and Britain's balance of payments. The Treasury was reluctant to transfer hard-won dollars to the United States for an exploration gamble that might well be unsuccessful. Shortage of dollars had been one of the factors in dissuading BP from entering the high-cost oil play in the Cook Inlet. But the exploration programme itself in such a remote area was far from cheap, and it was here that the Kern Oil properties in California which BP had bought in 1957 came into their own. These wells, discovered and developed in the mid-1920s, were close to Bakersfield in 'grapes of wrath' country. When BP executives first went over to see them they were horrified. The oil did not flow under its own pressure, as in the Middle East, but had to be pumped out at a rate of between five and ten barrels a day for each well. This was not unusual for the United States, but it came as a shock to oilmen accustomed virtually to turning a tap to produce oil. The method of pumping was even more archaic. For each group of half a dozen wells a steam engine drove a huge flat wheel from which bits of wire were attached to individual wells. As the wheel went round the wires would jerk the pump at each well in succession and a glug of oil would come out and trickle into a gutter dug in the ground. The gutters ran into a dammed-up waddy from where once a day a truck would come and pump out the oil that had collected. It was thick, almost solid, but was particularly useful in California for making asphalt, selling at a profit of about a dollar a barrel. Production had declined nearly to zero and BP seriously considered disposing of the properties. Then the Tidewater Oil Company, which also had leases in the Kern River field, became interested in the possibilities of secondary recovery which could greatly increase production. An agreement was negotiated and Tidewater took over operation of the field in return for paying BP a share of the profits. Production was increased to the point where BP was getting between two and three million dollars a year—and it was this money that was used to finance the Alaskan exploration. An added benefit was that it gave BP the 'fifty-cent dollar', so-called because oil companies with a US tax liability can write-off against profits any expenses involved in oil exploration as well as a depletion allowance for any oil discovered, effectively halving exploration costs. This applies to American oil companies in whatever part of the world they operate and has been a major factor in their rapid growth as compared with British companies, who have to capitalise their expenses and write them off over ten years.

And so the decision was taken at BP's head office in London to drill on the North Slope, a major step compared with the relatively inexpensive surveys that had gone before, but ultimately the only way of finding out if oil was there or not. It meant spending some three million dollars a year to begin with out of BP's total world-wide exploration budget of some fifty million dollars. Most of this was committed to maintaining production in countries already producing oil, leaving about ten million dollars for wildcatting in new areas. At that time, for instance, BP was searching for oil in Nigeria, Libya, Colombia, Canada and Papua. The first two areas paid off handsomely, Colombia was fairly successful, Canada barely paid for itself in exploration terms, and Papua was a complete failure. And so might the North Slope have been, for the geologists had made a mistake about the foothill folds north of the Brooks Range. They looked promising, but they did not go down to any depth. For a while it seemed that the whole venture had been a wasted effort.

In the summer of 1963 BP barged a Canadian drilling rig 2,000 miles down the Mackenzie River to the Arctic Ocean, then along the coast of the North Slope and into the Colville River to a landing point at Pingo Beach. There the barges were unloaded, the sections of the rig put together, and as winter conditions enveloped the North Slope and the ground hardened into a protective covering of ice and snow, the complex operation began of hauling the whole outfit sixty miles across country to the drilling location that had been selected in the Umiat area. It was thought that BP would be the first oil company to drill on the North Slope since the US Navy had completed its exploration programme in 1953, but in fact Colorado Oil and Gas beat them to it. Using a small rig that the US Navy had left behind, they spudded-in a shallow well one mile from the Gubik gasfield in July of that year. It was only drilled to 2,000 feet, however, as a dry hole, and the company showed no more interest in the area.

Between December 1963 and early 1965 BP's Canadian crews drilled six wells in the Umiat area under harsh and unfamiliar conditions. Blizzards and temperatures of 60 degrees below zero made life almost unbearable for the men, even Canadians used to the cold. Each well first had to penetrate a thousand feet of frozen ground where temperatures were so low that steel equipment fractured and normal lubricants solidified. And all to no purpose, for the reservoirs at depth were too thin and the small amounts of oil and gas found were of no economic value.

'We wasted money drilling in the foothills,' explains one of BP's geologists. 'If we'd studied the US Naval reports more closely we might not have gone to that area, for they showed that the prospects there were not all that good.'

While the drilling was going on, however, BP and Sinclair were carrying out surveys to the north along the coastal plain of the North Slope. These were all seismic surveys, for there were no rocks or outcrops available for geological examination. It had been known from the US Navy's drilling that oil-bearing sands existed at Point Barrow; the question was, did they continue all along the coastal belt? In 1963 a brilliant young Scotsman, Jim Spence, had come out to BP's Los Angeles office as chief geologist for the Alaskan operation. As a result of seismic surveys in the winter of 1963–4, he and the exploration team defined for the first time two arches on the coast, an enormous buried dome near the Colville River Delta and a smaller faulted structure at Prudhoe Bay, further to the east. It was felt that Prudhoe was more likely to contain oil than Colville, because of a tilt and then a reverse tilt which had led to the oil migrating eastwards. But both were good prospects and in September 1964 BP approached the state government, requesting that all the state selected lands covering Colville and Prudhoe be put up for competitive bidding. At the end of the year the state of Alaska made its first offer of North Slope acreage for leasing but decided to put up only half of the area requested and then chose Colville rather than Prudhoe. Since no one else was very interested, BP and Sinclair secured 318,000 acres over the whole Colville dome for an average of less than six dollars an acre. Had Prudhoe Bay been put up first, and it was only a matter of chance that it was not, the two companies might have acquired the whole of that structure for very little more cost.

In July 1965, while BP was making plans to bring the Canadian rig across from Umiat to drill at Colville, the state put up the first acreage at Prudhoe Bay for leasing. By this time, although the American companies were still mainly interested in the Cook Inlet operations, they had begun to take more notice of what BP and Sinclair were up to on the North Slope, especially Richfield, Humble Oil, Atlantic Refining and Pan American, who had all carried out geophysical surveys of their own in the area. Humble and Richfield had, in fact, joined forces to share the acreage they had both acquired in previous federal land sales and were waiting to see the results of BP's drilling programme. Meanwhile the federal government, which still owned most of the North Slope except the two million acres along the coast that had been selected by the state, was planning further lease sales.

Just before the Prudhoe lease sale, Sinclair decided to withdraw from the project. It had never been greatly in favour of going to the North Slope, persuded more by BP's better-than-average record of oil discovery than anything else. Now, after the total failure of the drilling operation in the Brooks Range foothills, the company had had enough. Without Sinclair's support, and in spite of a very limited dollar allocation, BP went ahead on its own, bidding against the far greater resources of the American companies. Deciding there was a chance that the oil-bearing sands were thin on top of the structure and thicker on the flanks, the British company spread its bids to include those leases which seemed to be around the outer edge, acquiring thirty-two blocks for $1,440,000 or an average of just over sixteen dollars an acre. Richfield-Humble went for the forty square miles crest of the structure, some of which it acquired for ninety-three dollars an acre against bids of forty-seven dollars by BP, twelve dollars by Mobil and Phillips combined, and six dollars by Atlantic Refining.

Hopes were again dashed early in 1966 when the well drilled on the crest of the Colville structure found only small quantities of oil, the main accumulation having apparently migrated elsewhere. It was BP's seventh well on the North Slope, and all of them had proved a bitter disappointment. With dollars running even shorter and Sinclair wanting nothing more to do with the operation, BP sought for a new partner and eventually came to an agreement with Union Oil who began drilling a second well, Kookpuk No. 1, on farmed-out acreage in the central part of Colville. Meanwhile Atlantic Refining, which had bid so low for Prudhoe acreage, had regained its position on the North Slope by merging at the end of 1965 with the Richfield Corporation. The New Atlantic Richfield Oil Company, still in partnership with Humble, decided to go ahead with the drilling of a wildcat well that had already been planned at a location between the Tooklik and Sagavanirktok rivers. This was Susie Unit No. 1, which began drilling in March 1966.

In January 1967 a further competitive sale of Prudhoe Bay leases was held by the state, only 38,000 acres this time, compared with the 403,000 acres before. But with Atlantic Richfield now drilling on the

North Slope as well as BP, the added interest in the area was obviously going to push up the price. Sinclair and Union Oil both declined to come in with BP and the company once more had to bid alone, with only a meagre $250,000 available. By carefully spreading the dollars out and going for blocks where there was no real chance of outbidding by competitors, BP secured further acreage down-flank from the crest of the structure. Six blocks were obtained for $44,000 a block, an average of seventeen dollars an acre compared with the overall average for the sale of thirty-five dollars an acre. Even that was trifling compared with the prices paid for leases in 1969, after oil had been discovered, when similar areas went for up to $28,000 an acre. By refusing to continue in the operation with BP, Sinclair lost the opportunity of having a lion's share of the biggest oil find ever made in the United States. The two companies had more geological information of the area than anyone else and with Sinclair's financial support they could probably have captured almost the entire Prudhoe Bay structure. As it happened, BP was wrong about the oil-bearing sands on the crest of the structure. They were thicker than had been thought. But then it turned out that they contained a large proportion of natural gas. Most of the oil lay in the flanks—and BP ultimately found itself with more than 60 per cent of the oil reserves of the structure. It was right for the wrong reasons. For Sinclair it was the last chance of staying in business as an independent company. In 1968 it was taken over by Atlantic Richfield. Had Sinclair stayed with the gamble, the situation might well have been reversed with Sinclair taking over Atlantic Richfield.

But the chances of the gamble coming off seemed even slimmer when, shortly after the lease sale, Atlantic Richfield's Susie well turned out to be dry and a few weeks later the second Colville well, being drilled by Union Oil, was also a failure. At that point even BP's determination failed. Over a period of five years it had drilled or been associated with eight of the ten wells on the North Slope and spent over thirty million dollars with nothing to show for it. BP could not afford to go on any longer and decided to cut its losses and leave. The drilling rig was stacked at Pingo Beach ready for shipment out. The Los Angeles office was closed down, its staff disbanded, the scientific records filed away in store. It seemed like just another of the failures which more often than not are the result of wildcat oil exploration.

And then in March 1968 came the announcement that Atlantic Richfield's last chance well at Prudhoe had struck significant quantities of oil and natural gas. In June, using the Canadian rig that BP had relinquished and which had been stacked at Pingo Beach, the company found oil in the same formation in a second well at Sag River, seven miles to the south-west. This indicated a large oilfield, confirmed shortly afterwards by an independent estimate that the field contained between five and ten billion barrels of recoverable oil, more than any other single find in the history of the United States. The announcement electrified the oil world, starting one of the biggest oil rushes ever as companies with leases in the area scrambled to bring in rigs by every conceivable route in preparation for the winter drilling season.

Rejecting offers by Atlantic Richfield to take over all the Prudhoe acreage, BP decided to return to carry out its own drilling operations. It meant a frantic rush to find a rig and assemble all the necessary equipment. With no roads to the North Slope—or even on the Slope itself at that time—there were only three possible routes; to fly everything in from Fairbanks, which was prohibitively expensive, or to ship by barge along the Mackenzie River from Canada or by way of the Aleutians and the Bering Straits. In either of the two sea routes there were only six short weeks of summer when the pack-ice in the Arctic Ocean was sufficiently broken up to allow barges through. It was too late in any case to use the Mackenzie River route which BP had pioneered before, and no one had ever tried such a large commercial shipment through the Bering Straits. But that was the only possible way.

Garth Curtis, one of BP's world-roving troubleshooters, was brought over from Das Island in the Persian Gulf to take charge of the operation. He arrived in Anchorage at the beginning of July, his first visit to Alaska, his first experience of Arctic conditions. In the course of one week he saw sixty rigs that were available for sale or hire. Eventually he chose one from the far side of Cook Inlet that had been drilling on the Tyonek Indian reservation. It was hauled through the forest, broken down into sections and loaded on to two barges, 4,500 tons together with all the equipment and supplies. The barges set sail on the long voyage round the Aleutian Islands, northwards through the Bering Straits, and then eastwards through the Arctic Ocean past Point Barrow, eventually arriving at the proposed depot on Foggy Island, a few miles to the east of Prudhoe Bay, in mid-August, just in time before the pack-ice started to re-form. The island lived up to its name, for visibility dropped to zero as the barges approached the landing point and they had to nose their way in to the shore using radar. The total cost of the operation, even before drilling started was five million pounds.

Once at Foggy Island they had to wait until the winter freeze-up, for the ground was too marshy for vehicles to transport the rig and equipment to the selected drilling site. The freeze came late in 1968.

It was not until mid-November that the barges could be unloaded and the rig hauled over the ice-frozen ground to the banks of the small Putuligayuk River, so named by the Eskimos, five miles from the coast and three miles south of Atlantic Richfield's original discovery well. Drilling began on November 20; meanwhile plans were made to charter two Hercules transport planes for $250,000 each a month to air-freight in two more complete drilling rigs. In March 1969 the Put River No. 1 well found oil in large quantities and BP knew for sure that it had a major share of the Prudhoe Bay oilfield, its own reserves alone confirmed by later drilling to be in the region of five billion barrels of recoverable oil, worth something like fifteen billion dollars excluding the cost of getting it out. Those costs were inevitably going to be very high, however. Even if BP could have raised the money in Europe the UK balance of payments situation made it impossible for it to be transferred into dollars. The operation would have to be financed largely within the United States. Also, BP had little experience of the very different industrial conditions in the United States, more closely regulated by government authorities at all levels than anywhere else in the world. Just as when the company had first embarked on its venture in America by going into partnership with Sinclair, the answer again seemed to be in finding an American partner with whom to develop Alaskan oil. Only this time BP would be negotiating from a position of great strength, suddenly finding itself the owner of a significant proportion of total US domestic reserves—perhaps as much as 12 per cent —at just the time when the country needed them most. BP had more than just a foothold in the American market. It was now a major operator.

No one received the news about the Alaskan discovery with greater delight than the man who had taken over in January 1969 as the chairman of BP, Sir Eric Drake (he was knighted in 1970). His arrival in the most august office of the company's executive suite, thirty-one floors up in BP's new glass and steel London headquarters, with its richly worn Persian carpet and famous Lutyens octagonal table as reminders of things past, heralded a new era for BP. Going into the United States and thus achieving for BP the status of a truly international oil company had long been a dream of his, ever since the traumatic days of 1951 when he, as general manager in Iran at that time, suffered the greatest indignity of anyone at being thrown out of the country, and then the years between 1952 and 1954 when as BP's representative in the United States he became schooled in the complexities of American oil politics. Even before BP had actually found oil at Prudhoe Bay, and while he was still deputy chairman, he had taken one of the most far-reaching decisions in BP's history—and one which would have cost the company $300 million had something gone wrong and oil not been found.

In the third week of November 1968, while BP was still going through a list of potential partners in the United States with whom it could merge to develop North Slope oil, its former partner Sinclair, sadly regretting the day it had backed out of the Alaskan operation, became the subject of a takeover bid by the financial conglomerate of Gulf and Western. It was the final insult to the wounded pride of the Sinclair management. Hoping to fend off the bid, they proposed a merger with Atlantic Richfield. It was soon evident that this would be opposed by the US Justice Department on the grounds of anti-trust, since both companies had large marketing networks in the New England states. Atlantic suggested that BP should acquire Sinclair's marketing and refining facilities in those states for $300 million, to be paid in instalments once production began from Prudhoe Bay. It was this 'hire-purchase' offer that finally convinced the doubters in BP that its share of the reserves must be very large, since Atlantic had by then drilled two wells at Prudhoe and knew more about the field than anyone else. The offer, made on a Monday, was contingent on completion of a deal by the end of that week to meet a timetable set by Atlantic's corporate requirements. The chairman of BP, then Sir Maurice Bridgeman, was away ill and the decision rested with the deputy chairman, Sir Eric Drake.

'I took the chance and went ahead with the deal,' Sir Eric recalls. 'It was practically certain that we had oil on our leases, but it hadn't been proved. They were a rather anxious three months, until our well at Put River was brought in. If it hadn't been successful, it would have cost us $300 million.'

Although one suspects that there was a get-out clause somewhere in the small print, it was not a risk that one would have expected an accountant to take. Which is what Sir Eric's profession is, his career with BP starting in the accountant's department in 1935, going through thirteen years in the Middle East and two in America, then back to Britain to head the newly formed supply and development division and to nurse the expanding tanker fleet through its growing pains, finally joining the board in 1957 at the age of forty-six as one of its youngest members, becoming deputy chairman five years later. But underneath Sir Eric's urbane, friendly but slightly austere outward manner there is a toughness that those he has negotiated with have had cause to respect, and an unconventional way of doing things not expected of him that has even surprised his colleagues sometimes. A case in point was his action over the Sohio deal.

The Standard Oil Company of Ohio, founded as the original nucleus of the old Standard Oil Trust before it was broken up, was the company chosen by BP for a possible merger. The Sinclair properties on the East Coast had given BP a marketing outlet in the United States for the first time—and also permitted the Atlantic-Sinclair merger to go through—but they did not provide the financial and top-class management resources that BP needed in America. Discussions with Sohio began in earnest in March 1969. The deal that emerged called for an amalgamation with Sohio of BP's former-Sinclair eastern marketing facilities and its Alaskan production properties, in return for an initial 25 per cent shareholding in the American company. A dividend would not be payable until January 1975, or when North Slope production reached 200,000 barrels a day, whichever came first. As production increased, so would BP's shareholding in Sohio, rising to a controlling 51 per cent at 450,000 barrels a day and a maximum of 54 per cent if 600,000 barrels a day was reached by January 1978. The whole operation would be managed by Sohio which would market under its existing brands and also those of BP in the former Sinclair area, but BP Exploration Alaska would continue to operate the Alaskan properties on behalf of the amalgamated company. As Sir Eric admitted, the whole deal was a 'huge calculated risk'—until it was finally proved that the Prudhoe Bay field could produce at a total rate of two million barrels a day, which gave BP the quantities required under the sliding-scale arrangement. If those production rates were not met for any reason by 1978, BP's shareholding in Sohio would remain at whatever point it had then reached and BP would retain any extra oil produced thereafter.

There were many legal problems to be solved. BP had to conform to the requirements of the Securities and Exchange Commission in order to set up the BP Oil Corporation and to have its shares quoted on American stock markets. It had to comb through its many thousands of individual shareholders all over the world, including the 3 per cent Iranian shareholders, to make sure they were nationals of countries giving reciprocal leaseholding treatment to US citizens; if more than a certain small percentage were not, if for instance there had been a number of Soviet shareholders, then BP could not operate directly in the United States. Once these hurdles had been overcome, the only legal discrimination against foreigners was that a company could not own US flag ships if more than 20 per cent of its shareholders were not US citizens—and by law any oil moved from one point in the United States to another had to be by US tanker. This restriction applied to BP, of course, but also to a number of American companies, including Sohio itself and Shell of America. The biggest problems of all were the American anti-trust laws which dated back to the old Standard Oil Trust and the robber-baron days. It was no good asking lawyers beforehand, because although they did in fact give as their opinion that the deal was in order, they could not guarantee this. American laws and guidelines in this respect are intentionally vague, the idea being that the US Attorney General should have the right to regulate matters according to the needs of the day. And so it was, just when everything seemed straightforward and the merger plans of the two companies were going ahead, that towards the end of 1969 the Justice Department stepped in and announced that the deal was illegal on the grounds of anti-trust.

There was an outcry of criticism in Britain and Europe, where it was felt that the United States was discriminating against a foreign company trying to operate there at a time when American investment in European countries was greater than ever. It was perhaps forgotten that the balance still remained in Europe's favour, the figures suggesting that while Americans held about $20,000 million of long-term capital in Western Europe, Europeans held some $25,000 million in the United States. In Britain's case the balance was even more favourable. If the deal went through, Sohio would move up from fifteenth to at least twelfth place among US oil companies and join the many major American concerns which had British parents, such as Shell, Bowater, British-American Tobacco and EMI. The US Justice Department had taken the same line over the question of mergers between American companies, but that did not prevent the BP–Sohio deal threatening to become a major international incident, involving not just the companies but the governments of the two countries as well.

Arguments between the lawyers of the Justice Department and the two companies could have gone on for months, years in fact, if the case had reached the courts, where such a thing as a quick settlement is virtually impossible under the American legal system. But in the middle of all the delicate negotiations Sir Eric Drake, the quiet, seemingly withdrawn accountant, surprised everyone by making a sudden flying visit to Washington and personally confronting the Attorney General to 'call his bluff' by putting the problem plainly: 'If we aren't allowed to do this, then tell us what we can do and we'll go along with whatever you say.' It was almost unheard of in America for the chairman of a major company to take such a direct step instead of leaving it all to the lawyers. It caught the Justice Department off balance and the upshot was that BP probably got better terms in the end than an American company could have expected. As a result of the meeting between the two men, it was agreed that BP and Sohio would sell certain of their service stations where these overlapped in particular areas so that their combined outlets would not command an undue share of the market. With that done, the deal was given the Justice Department's blessing and was finally completed on New Year's Day, 1970. BP at last had the major stake in America that had for so long been its ambition. And the British Treasury could take heart from the fact that its shareholding in BP now stood to be worth considerably more than the country's entire gold reserves.

North Slope oil is still not flowing, the whole operation blocked by problems that were to come later and which are dealt with in following chapters. But whatever happens, as the United States faces a growing energy crisis caused by a shortage of oil, there is no doubt that it will be produced sometime. It is Sir Eric Drake's personal ambition to see Alaskan oil begin to flow before he retires. It would be the crowning achievement of his years of stewardship as BP's chairman.

II

Oil Boom

Once, it was the gold rush that lured the adventurous with dreams of untold wealth. Today it is the oil boom, conjuring up visions of globe-jetting Texan millionaires, mysterious Greeks, Cadillacs in the desert, the tinsel palaces of Arab sheikhs. It is the idea that catches the imagination, for, unlike gold, oil is never seen at any point on its journey from thousands of feet beneath the ground to the gasoline tank of an automobile or the boiler of a central heating unit—unless by accident when it is washed up on holiday beaches as an evil-smelling pollutant, somewhat tarnishing the image of the stuff that has been called 'black gold'. But that is the aftermath, by which time the boom conditions have settled down to the routine business of oilfield development, refining and marketing. An oil boom is a peculiar thing in itself. For years the exploration companies poke about in remote corners of the world and little notice is taken of them. Then oil is discovered and suddenly it is being talked about by taxi drivers in Rome, bankers in Hong Kong, investment-conscious widows in New York. Fortunes are made overnight, men die in the frantic and dangerous haste to get a stake in it, old ways of living undergo radical change. All this happened in the year of the North Slope oil boom, from September 1968 to September 1969. The discovery was the biggest ever made in the United States, bigger even than the five-billion-barrel East Texas field found in 1930, with the irony, however, that not a drop of oil has yet been brought out of northern Alaska and is not likely to for some years to come.

What started the boom was Atlantic Richfield's second Prudhoe Bay discovery, which proved the existence of a large oilfield, and later independent estimates by DeGoyler and MacNaughton and top international oil consultant Walter J. Levy that if the first finds were matched elsewhere on the North Slope a projection of thirty to forty billion barrels of recoverable reserves 'would not be improbably large'. This is more than the rest of US known oil reserves put together, approaching the scale of the Middle East giants—Kuwait's sixty-two-billion-barrel Burgan field, for instance. Wilder estimates in the Alaska oil reserves guessing game put the figure as high as 100 billion barrels and over forty trillion cubic feet of non-associated natural gas. In the summer and fall of 1968 oil industry shares shot up on the world stock exchanges, especially those of such companies as Atlantic Richfield and BP who had considerable acreage at Prudhoe, and in some cases doubled over a period of a few weeks. What gave the boom its intensity was the considerable amount of acreage around Prudhoe Bay that still remained to be leased and the decision by the state of Alaska to hold a sale of those leases on September 10, 1969. The companies had just one year to carry out exploration work and decide what those leases might be worth and what they would have to bid to get them. In the case of companies with leases already, this meant drilling wells to see if they were lucky winners in the oil-game stakes. Those without had to content themselves with carrying out seismic surveys to make up lost ground in their geological knowledge of the area.

Until 1968 BP and Atlantic Richfield had drilled their wells only in winter, closing them down for the summer months if they were not completed by the time the ice broke up. Winter might have been tough on the men, but it was the ideal time operationally, when vehicles could move easily around on the frozen-hard ground, planes could land anywhere on the flat surface of the Slope, and rigs could be hauled by tractor from one site to another. During the summer months, when the North Slope becomes a marshland, efforts were concentrated on bringing in bulk supplies and equipment, the biggest problem of all in such a remote area. Everything had to be supplied from outside, from complete drilling rigs to food and medical supplies. Since there was no road at that time north of Fairbanks, there were only three possible routes. Two were by sea—up through Canada by way of the Mackenzie River and westwards along the Arctic coast to Prudhoe Bay, or through the Bering Straits from Seattle and then eastwards to Prudhoe. The alternative was by rail to Fairbanks and then by air freight the last 390 miles to the North Slope. As the cost of shipping was between $90 to $125 a ton, compared with about $270 a ton by air, the sea routes were the cheapest and most logical. But careful planning was required since there were only a few weeks in midsummer when barges could get through the ocean pack ice. Because of the difficulty in deciding just what supplies would be needed and in what quantity, there was a tendency to over-stock. On the other hand, any items forgotten or urgently required later had to be brought in by the more expensive air route.

A typical rig, for instance, weighing 500 tons, would require 175 tons of steel drill-pipe and casing and another 150 tons of mud, chemicals and other materials. It would need a prefabricated camp to accommodate up to forty-five men complete with kitchens, mess hall, living quarters and recreation room—another 200 tons. Then there would be tractors and service vehicles and diesel fuel to run both them and the rig, as much as 3,000 gallons a day at up to $1.08 a gallon with every engine kept running all the time in order to prevent it freezing up and a single truck eating up 100 gallons a day. A total weight of equipment and supplies of perhaps 1,500 tons for one self-contained drilling outfit. On top of that there were the wages of the crew—$25,000 a year for a driller, $20,000 for a labourer, $30,000 for a cat-skinner—the best paid of all—working twelve-hour shifts six weeks on and three weeks off; another fifty dollars a day per man for catering, each man needing some 4,000 calories daily because of the cold; and the cost of servicing the rig by plane. It was not unusual for a helicopter to burn 500 gallons of fuel in order to deliver 1,500 gallons. And then there were the sheer problems of living and working in temperatures of 60 degrees below. Metal became brittle and snapped. Drilling mud froze. It was a major problem to find water for drinking and washing—amid all that ice and snow, water-wells had to be drilled. The efficiency of a man was reduced by as much as 80 per cent. A normal five-hour job might take three days to complete; because of having to wear mittens and gloves the men could not use their fingers to tighten or unscrew bolts when working outside. The floor of a rig was usually heated and covered over with canvas and boarding to protect the crew from the constant winds blowing across the North Slope. But it was not much help to the man who had to stack the drill pipe at the top of the rig, perhaps the worst job of all.

'Every time we had to make a join it was a real adventure,' recalls Lee Wilson, drilling superintendent for Atlantic Richfield who had come to Alaska in 1964 and worked initially in the Cook Inlet before tackling the problems of the North Slope. 'When it got really cold there'd be breakages in the pipe and the blades of the cats. We had to pre-heat the casing before running it in. When we first started we found the permafrost under the surface washed out very quickly when we used fresh-water muds. The heat would melt the sides of the well and they'd just cave in. Then, while we were trying to fix that, the whole thing would freeze up again and we'd be back where we started. We found the best and cheapest way was to dig down for the first 100 feet without drilling or using circulating fluids.'

This 'dry bucket' technique had been used in other industries but not in oil drilling before. It was pioneered on the North Slope, one of the special methods needed to adapt to special conditions. Once the conductor pipe was in place drilling would continue using chilled brine circulating fluids so as not to melt the permafrost—while on the other hand the casing had to be heated before it was run in to prevent it being cracked by the intense cold.

Not surprisingly, the North Slope was the most expensive area ever drilled by the oil industry. Costs were estimated to be about $142 a foot compared with twelve dollars a foot in Texas. Running costs alone were $20,000 a day. Atlantic Richfield and Humble spent $4·5 million to drill the Susie No. 1 well—which turned out to be dry. A later well drilled near that site, Nora No. 1, also by the same company and also dry, was completed at a state-record depth of 17,658 feet and at a record cost of seven million dollars.

Once oil was discovered, however, time became more important than cost. Before, only some of the seismic crews had worked in summer. Now drilling too became a year-round operation. Atlantic Richfield was the first to attempt this. In the spring of 1968 the company hauled the Canadian rig at Pingo Beach ninety miles down the Colville River to the coast, then across the sea-ice and through Prudhoe Bay to the Sag River State No. 1 location. Since it was intended to drill through the summer, something had to be done to provide a working surface on the ground when it became marshy from melted ice. Gravel from river beds was the answer. Some 500,000 yards were taken from the Sagavanirktok River and spread in a pad five feet thick under where the rig was to be erected and in the area around where men and machines would be working. It was also used to make an airplane landing strip. The Sag River location, in fact, became the company's main base, with a permanent camp for 200 men and even a small refinery, completed in April 1969, producing 1,000 barrels a day of diesel oil at forty-five cents a gallon compared with over one dollar a gallon to ship it in. Any amounts over and above Atlantic Richfield's own requirements were sold to other companies in the area.

It was the availability of abundant supplies of gravel in the beds of lakes and rivers that made summer operations possible for there were no other materials handy that could have done the job. So much of it was eventually required that shipping it in would have been out of the question. There were criticisms from conservationists about the effect its removal would have on fish coming up the rivers, and there was also concern about the damage that was being caused by churning up the tundra. The whole question of Arctic ecology came to be a vital factor, a major reason for holding up later development. But for the moment, in the summer and winter of 1968, all that mattered was the excitement of the oil boom and the furious activity that it engendered. In the land of the last frontier it was like the pioneering days all over again.

While Atlantic Richfield and BP concentrated on drilling around Prudhoe Bay to determine the size and boundaries of the oilfield, other companies brought in rigs to test their own leases. Mobil, as operator for itself and Phillips, barged in a rig and supplies by the Mackenzie River route. Other companies, including Pan American, Texaco, Standard of California and Colorado Oil and Gas, flew in rigs by Hercules aircraft from Fairbanks. By March 1969, the height of the winter drilling season, Mobil and BP had completed their first wells, six more were drilling, four were being rigged up, and locations had been staked out for a further four. Most of the wells were given Eskimo names from the rivers in the area, such as Kavik, Toolik and Shaviovik. They were 'tight' wells, meaning that little or no information about them was released in view of the pending lease sale. The competition for leases was obviously going to be fierce and the advantage would go to those companies with the most complete geological knowledge.

The North Slope had changed out of all recognition. A jetty had been built on the shores of Prudhoe Bay for bringing in the summer shipments. Apart from the rigs fingering up from the once desolate wilderness, no fewer than seven landing strips had been built within a seventy-mile radius. One of these, Standard of California's Deadhorse field, was eventually taken over by the state and became the main airfield for the region. Companies providing service and supplies for the oil industry began moving in, setting up workshops and warehouses, stacking up huge piles of steel pipe and casing, bags of cement, chemicals for drilling muds. And fuel drums—everywhere barrels of diesel fuel without which the whole operation would have ground to a halt. Gravel roads snaked in all directions linking base camps with the drilling sites. Nearly 2,000 men were working in the area at the peak of the season. There were vehicles of every kind, from small pick-ups to caterpillar tractors, road graders, and massive overland transporters. For as an additional means of bringing supplies to the North Slope the Alaskan state government had driven a 330-mile winter road over the

Brooks Mountains from Livengood, a town at the end of a seventy-five-mile highway extending north from Fairbanks. The road was in affect a trail bulldozed over the ice-hardened ground, cutting through the forested valleys of central Alaska, passing old gold-mining camps like Wiseman and Coldfoot, then across the Anaktuvuk Pass of the Brooks Range and down on to the North Slope. It was named the Hickel Highway after the then governor of the state, Walter J. Hickel, who gave permission for its construction—and it immediately caused a storm of controversy, incensing those who were concerned about conservation. Huge convoys made the twelve-day journey with thousands of tons of supplies, the tractor-drawn sleds crawling along at a few miles an hour, the drivers taking over in turn and sleeping in the special quarters provided since it was a round-the-clock operation. But in practice there was little saving by using this new route. Haulage costs of up to $240 a ton were not much less than the normal air freight from Fairbanks.

It was in providing the oil companies with transportation for their supplies, either by aircraft or road haulage, that Alaskan industry benefited most from the oil boom. Most of the drillers and their crews came from outside—from Canada or the oilfields of Texas, Oklahoma and California. Outside companies with long experience in the oil business provided most of the specialised services required, from contracting to drill a complete well to testing results by electronic well-logging, supplying chemicals, drilling fluids and drill-pipe, and advising on the many complex problems encountered. Oil-well servicing is big business in itself and such firms mushroom wherever an important strike is made. But transportation in the harsh and dangerous conditions of the northland, particularly by air, was something that required Alaskan experience and know-how. The centre for this massive exercise in logistics was Fairbanks, the 'Golden Heart of Alaska' city. (Anchorage is the 'All American' city.)

Fairbanks had for years been in a state of economic eclipse. Growing from a gold rush camp at the turn of the century, the population had declined from 11,000 to less than 2,000 by 1920 as miners moved on to other finds and the big mining companies that came in their wake discontinued operations when the gold became played out. Completion of the Alaska Railroad in 1923 helped to keep the city alive, then military activities during and after the Second World War stimulated growth to the point where the population was 37,000 by 1960, plus 10,000 military personnel. This figure remained static for the next eight years. Because of the extreme cold of the long winter months when most construction activity had to shut down, unemployment was high, averaging 8 per cent in the city and up to 33 per cent in some of the outlying districts. The extremes of temperature, ranging from 99 degrees above in summer to 66 below in winter, also brought problems of adjustment to new residents trying to cope with it. 'Cabin fever', the neurosis caused by being cooped up in a house during the long, dark winter months, is an often-heard expression in Fairbanks, the Alaskan equivalent of 'new town blues'. It has a parallel in the oil camps, when men have suddenly attacked others for no apparent reason. The Alaskan climate can have a strange effect on newcomers. One geologist on a survey party in summer, during the twenty-four hours of daylight, was unable to sleep until he had got hold of a large box that had previously contained toilet rolls, made a large hole in it like a headsman's block and smaller holes for ventilation, and then put his head into the darkness inside. 'His snores echoed like thunder in that damn box,' a colleague recalls with some feeling.

Fairbanks, like Anchorage, has had its share of natural catastrophe in recent years. In Anchorage disaster came at 5.30 pm on March 27, 1964—Good Friday—during the earthquake which struck a large area of southern Alaska and virtually demolished the seaport of Valdez. A whole street in the centre of the city was engulfed, over 100 people killed, and many houses on the outskirts destroyed by huge rock-slides. Everyone has a story of this biggest earthquake ever recorded in North America—the secretary on the top floor of a new office building who escaped after several storeys had sunk into the chasm, the man who went home early because of his uneasiness that all the stray dogs and birds seemed to have disappeared from the streets, only to be injured when his house partially collapsed. In Fairbanks it was a flood that brought disaster. During mid-August 1967 the Chena River rose eighteen feet above its normal level, inundating most of the city, driving thousands from their homes and causing over $200 million of damage. But like the people of Anchorage the residents rolled up their sleeves and, helped by federal grants and loans, got down to the job of rebuilding their city.

The oil boom came at a time when the economic shot in the arm resulting from this disaster had waned and Fairbanks again faced the spectre of high unemployment and an outward migration from a lack of job opportunities. But within a matter of weeks Fairbanks airport had gone from a deficit operation to the state's most active airport in freight operations as the huge airlift to the North Slope got under way. Nearly 100 companies in the oil industry or related to it opened offices in the area, many of them leasing warehouse space as well. A crash programme was initiated in an attempt to solve the housing shortage, made more acute by a population estimated to have grown to 46,000 by 1969. Local businesses boomed—and not least of all the saloons and dance-halls so characteristic of Fairbanks, where oil drillers with their pockets full of dollars inflated prices to the point where a half-bottle of champagne and a girl to help you drink it cost twenty dollars.

The various Alaskan airline companies, such as Wien Consolidated, Alaska Airlines and Interior Airways, had mostly been formed by intrepid bush pilots of the early days, when a man would beg or borrow the down payment on a two-seater plane and gradually built up a charter business carrying hunters, fishermen and prospectors. There were very few proper airstrips, and most of the time when flying in the Arctic the pilots had to land on frozen lakes. The greatest danger in Alaskan flying was literally starving to death if a forced landing had to be made in the wilds. Air transportation developed rapidly until today even the smallest villages are included on some kind of scheduled air service. Jet passenger planes landing at a remote Eskimo settlement are no more uncommon than a country bus stopping in a small English village. But there are still a few bush pilots left to carry on the old tradition. Such as Don Sheldon, who operates a one-man charter company from the village of Talkeetna, 160 miles north of Anchorage, with the help of his wife Roberta, herself the daughter of one of the most famous old-time glacier pilots, Bob Reeves.

Don Sheldon came to Alaska from his home state of Wyoming in 1938, when he was seventeen. He got his pilot's licence in 1942 and has been flying ever since—in B.17 bombers during the war, then in Mexico for a few years, and continuously in Alaska after returning in 1948. Since then he has clocked up over 15,000 hours as a bush pilot, which is a lot of time for small planes such as Piper Cubs and Cessnas. He has specialised in flying scientific and climbing expeditions to the mountains of the Alaska Range, particularly the 20,320-foot Mount McKinley, the highest in North America, where winds can rage at 150 miles an hour in winter, temperatures fall to minus 60°F, and the chill factor can add five degrees of frost for every thousand feet climbed. One expedition he remembers particularly was in 1962, led by a wealthy Italian climber, Ricardo Kasin, who bet his personal fortune that he could climb the south face of the mountain without ever having seen it before. Sheldon landed the party on a glacier at 7,400 feet. They began the climb, then at 16,000 feet they were caught in a tremendous storm. For three days and nights they swung on their ropes against the sheer face of the mountain, unable to go up or down, while their hands, feet and faces froze. When the storm died down, although suffering severely from frostbite, they continued to struggle up and became the first crew to climb the south face. Sheldon rescued them after landing as high up as he dared—above the 9,000-foot level—which was itself a considerable feat. Landing on a glacier can be a hairy business, as BP's Charles Towill found when Sheldon landed him and a photographer on Mount McKinley early in 1969. There was a deeper layer of snow on the glacier that Sheldon had reckoned, and the Cessna ended up with one wing buried and the other pointing up at the sky. It took most of the day for them to tramp down a large enough area to make a runway and even then they hit a flurry of snow before taking off, just yards from a drop of several thousand feet, so that the windscreen became completely covered over. 'I think we'll make it,' was Sheldon's only comment as, his vision obscured and unable to see when they came to the end of the glacier, he eased back on the controls. They became airborne with only inches to spare.

It is a matter of pride with Sheldon that he has never had to send out a rescue call, although many times he has rescued others. But he has had his moments. He once took up a party of priests and when they were about to land he saw that one of the floats had become loose and was sticking up at the back. He turned to one of his passengers: 'Father, would you mind? Just tie this rope round your waist, then get out and stand on the back of the float to keep it flat.' The priest did so, although in little doubt that his time had come. But they landed safely with the priest standing on the float, his eyes screwed tight and shouting his prayers to the wind.

Sheldon recalls such incidents with the laconic humour and unconcern of all bush pilots. The vast interior and northern area of Alaska is his backyard; he knows each trapper's cabin and most of the Eskimos and Indians in the scattered settlements. The oil boom meant new business at a time when he was threatened by the expanding regular air services. In March 1969, for instance, he was commissioned to lay a string line over forty-two miles of the route of the proposed trans-Alaska pipeline, from near Stevens Village to Livengood, so that the land survey party would know where they had to bulldoze their track through. It would have taken over three weeks to mark the route on foot; Sheldon did it in one morning by flying low over the ground in a Piper Cub and stringing out the line behind him. 'A safe height was about three or four feet above the tree-tops,' he commented. Within eight days, bulldozers had followed the line and driven a track right through.

To the bigger Alaskan airline companies the oil boom meant an immediate search for any kind of transport aircraft with which to freight in supplies. By the end of 1968 there were eleven Lockheed

Hercules aircraft alone operating from Fairbanks to the North Slope, carrying 48,000 lb loads at around $4,500 a trip. Fairbanks airport swarmed with an amazing vareity of planes, including Dakotas and old wartime bombers, C46s, DC6s, F27s, Beavers, Otters, King Airs and 737s, as one of the biggest civilian airlifts in history got under way. Over 135,000 tons of freight were flown in during the twelve-month period and around 50,000 passengers carried to more than twenty different drilling sites. There were just not enough aircraft to go round and some oil companies had to resort to chartering their own, mostly DC7s. On the North Slope itself the most expensive operation was Atlantic Richfield's use of a giant Sikorsky Skycrane helicopter to move entire rigs, broken down into twenty-ton loads, from one site to another. Three rigs were moved in this way and kept supplied during a continuous operation in the summer of 1969—at $750 an hour flying time and more than $100,000 a month for stand-by. One seven-mile rig move was made in fifty-four hours' flying time. Even before the lease sale in September 1969 it was estimated that the oil industry had spent over $1,500 million on North Slope operations, compared with seventy million dollars on the ten dry holes drilled there from 1958 to 1968.

But this furious activity had taken its toll. Thirty-two men were killed on the North Slope during 1969, violent deaths which are a part of the oil game that the customer seldom gets to know about. Most of these were in the ten aircraft crashes that occurred, particularly when a Hercules transporter was lost with all its crew. Three were killed when a hovercraft crashed while being used on an experimental operation. Two men walked into helicopter blades. Another was crushed between a fork-lift truck and its load. Two men were drowned in the cement they were pouring to erect a building, one was killed in an explosion during a seismic survey. In addition there were many serious accidents amongst the drilling crews and the Arctic cold took a further toll in the way of snow blindness and amputations as a result of frostbite. Drilling for oil is always dangerous; the weather conditions on the North Slope make it that much more of a hazard.

As drilling gathered pace through the summer of 1969, most of it within a ten-mile radius of Prudhoe Bay, the oil companies began to build up a clearer geological picture of what they had found. The whole field lies within an area roughly forty-five miles long from west to east along the coast and eighteen miles wide, one of a series of broad anticlines underlying the North Slope and known collectively as the Barrow Arch. There are three main oil and gas bearing reservoirs in formations of lower Cretaceous to Triassic age, between 120–220 million years old, sealed by a stratigraphic fault of cap-rock plunging from west to east and partly underlying each other. The shallowest is the Kuparuk River sandstone, some six miles to the northwest of Prudhoe Bay. It is up to 800 feet thick, below a depth of 6,765 feet, and extends over an area of about 128,000 acres. Next is the most important and biggest formation, the multipool Prudhoe Bay group, formed in undulating sandstone up to 600 feet thick below a depth of 8,110 feet and covering some 368,640 acres. Slightly to the east is the deepest reservoir, formed in what has been classified as Lisburne limestone, up to 1,700 feet thick below 8,758 feet and covering 181,750 acres. Although by far the thickest of the reservoir rocks, it has a low permeability compared with the other two, whose rock is very porous and fine to medium grained, so that production rates from the Lisburne zone will be relatively smaller.

Of the twenty-three wells that had been completed by the end of the summer—nine by BP, seven by Arco-Humble, four by Mobil-Phillips, two by Socal (Standard of California) and one by Hamilton Brothers—only two failed to find any oil or gas (one each of BP's and Arco's), although the rest would not all necessarily prove to be commercial. Seven wells encountered two of the reservoirs and one, Socal's Kavearak Point wildcat, found oil in all three reservoirs.

As the main operators, BP, Atlantic Richfield and Humble Oil began to plan for a unitisation scheme in order to develop efficiently and economically as a single unit. In the early days of the industry in America a discovery by one company would bring a host of others to the area, each seeking to get as much oil out before the field dried up. Dozens of rigs would sprout up within a few yards of one another and the resulting over-production would depelete the field too quickly, and even leave most of the oil intact, for it has to be given time to seep steadily through the pores of the reservoir rock. The worst of these excesses were brought under control by government prorationing laws in such states as Texas, Louisiana and Oklahoma, limiting the production allowed from an indivdual well as a percentage of the estimated reserves of the whole field. This establishes a maximum efficient rate for each well and even this is further limited, depending on market demand, in order not to create an over-supply of oil and thus conserve resources—and also, some argue, to maintain artificially high prices. Alaska has no prorationing laws but since all the oil land is owned either by the state or federal governments, and in fact very little acreage in the entire state is privately owned, drilling and production regulations can be imposed more directly. The fact that only a few big companies can afford the expense of drilling on the North Slope is itself an asset in terms of conservation; accustomed to taking the long-term view it is in their own interests to work together for the most efficient development of the field. At their suggestion the state government has established spacing regulations on the basis of one well to 640 acres for the main Prudhoe Bay sand pool and the Lisburne lime pool, and one to 160 acres for the shallower Kuparuk River sand pool discovered by Standard of California.

This means that some 400 development wells will be drilled to put the entire field on production, each capable of producing between 15,000 and 20,000 barrels a day—compared with an average production rate from all US wells of a mere fifteen barrels a day. North Slope costs will be further reduced by drilling between four and eight wells, deviated at an angle downwards from one large gravel-based drilling pad. It was estimated that the field would begin production at a rate of 500,000 barrels a day, when 500 million cubic feet of associated gas a day would also be produced, and increase to 2·2 million barrels a day towards the end of the 1970s, by which time it would be accounting for 17 per cent of total US output and helping to meet a demand of twenty million barrels a day. If further oil was found on the North Slope, as seemed likely, these production rates might be doubled.

In the first heady days of the oil boom it was thought that North Slope oil might be available for export to Western Europe and Japan, the two fastest-growing oil markets in the world and both almost entirely dependent on imports from other countries. The United States might again become a net exporter of oil, as had been the case in the great expansionist years earlier in the century, and it was suggested that Alaskan oil could lessen the stranglehold that the Middle East, with its volatile and unpredictable politics, had on America's allies. Japan is particularly vulnerable in this repsect, relying on Arab countries for the great bulk of her oil requirements and possessing little production there through national oil companies of her own as do the Europeans. There had for a long time been very close connections between Japan and Alaska, situated as they are on either side of the North Pacific. Japan currently buys over 85 per cent of Alaska's total exports and through investments going back over fifteen years controls a considerable proportion of Alaska's business. A number of Japanese companies began searching for oil in southern Alaska as soon as the Cook Inlet discoveries were made, mostly in partnership with American interests. Although all the oil discovered so far is required for the US West Coast markets, natural gas offered another possibility. Apart from supplying local needs in Anchorage, the gasfields found in the Cook Inlet either had to be closed down or, where gas was being produced in conjunction with oil, flared off from the top of the drilling rigs. The quantity was not sufficient to make it economically worth building a pipeline to any American market, unlike the much vaster amounts of gas found on the North Slope for which such a pipeline is planned. It was sorely needed in Japan, however, and as a result of an agreement reached in 1966 between American and Japanese companies special tankers were constructed to supply 50,000 million cubic feet a year of liquefied natural gas from the Cook Inlet area to Tokyo for use as town gas and to power electricity generators. Shipments by refrigerated tanker started in 1969 from special facilities built at Port Nikiski near Anchorage. Meanwhile, Japanese companies stepped up their search for oil in various regions of Alaska. And it seemed that the North Slope might provide the alternative source of oil for which Japan had been looking for so long.

But the realities of the situation soon became apparent. In spite of the size of the Prudhoe Bay field, consumption in the US is rising so fast and additions to domestic reserves in the traditional oil-producing states are continuing to decline to such an extent that the home market will need all the oil that could for the moment be produced from the North Slope. By 1980, even at the highest rates visualised of some five million barrels a day, North Slope oil would be meeting only 25 per cent of total US demand. Its entire proved reserves would be the equivalent of only two years of US consumption. Further large-scale discoveries on the North Slope and in other parts of Alaska might change the picture and even leave enough over for oil to be exported to Japan. For the moment, however, with production from other states falling in relation to demand, it seems likely that the United States will herself have to rely on imports to supply around half the amount of oil she needs.

The big problem facing the oil companies in 1968, after the North Slope discovery, was how to get the oil out. The markets open for it within the US that were already dependent on oil imports to the extent that domestic production did not, or could not, meet demand fell into three main areas: the East Coast, where oil was being imported primarily from Venezeula, North Africa and the Middle East; the Midwest, where Canadian oil was being brought by pipeline from the Alberta oilfields to the district around Chicago; and the West Coast, being supplied with oil from Canada, the Persian Gulf and Indonesia. There were only two possible ways of transporting North Slope oil: by sea, through the Bering Strait, or overland by pipeline. The first would mean opening up the legendary North-west Passage as a regular shipping route, for long a dream of merchant adventurers but one which had never been fulfilled. If it could be done, tankers or even oil-carrying submarines would most logically serve the East Coast markets. The second method could supply oil to either of the other two markets, through a 2,900-mile pipeline across Canada to Chicago and the Midwest, or through an 800-mile pipeline southwards right through Alaska to an ice-free terminal on the south coast where it could be transferred to tankers operating the year round and shipped down to the West Coast refineries. Natural gas, for which there were plans to market initially some 3,000 million cubic feet a day, would have to be transported all the way by gas pipeline taking a route through Canada, either through British Columbia and on down to Los Angeles or via the Mackenzie River Valley and through Alberta to tie in with existing lines serving the Midwest. But the most economic route for an oil pipeline was undoubtedly the one through Alaska, less than a third the distance of the Canadian route. It was therefore on the project to build a trans-Alaska pipeline that the companies concentrated their engineering studies. Even that was estimated to cost $900 million for a forty-eight-inch line capable of carrying two million barrels a day, and any serious delay would considerably add to that figure. A Canadian line would cost around $2,500 million, similar to the price projected for a gas pipeline.

Various estimates were made at the time as to the cost of North Slope oil. Professor Arlon Tussing of the Federal Field Committee for Development Planning in Alaska reckoned the wellhead cost, including discovery, development and production, to be between 24–54 cents a barrel based on recoverable reserves of 10,000 million barrels, compared with eighty cents–$1.25 a barrel in the Cook Inlet. Transportation through the trans-Alaska pipeline would add a further sixty cents a barrel. The oil companies, reluctant to divulge their estimates of production costs, would probably put these figures higher; on the other hand Professor M. A. Adelman, the respected oil economist, estimated costs to be even lower. Transportation by a Canadian pipeline, either the 1,700 miles to Edmonton and linking in with existing pipelines or the 2,900 miles direct to Chicago, would probably cost between eighty cents to a dollar a barrel. In addition, oil could also be sent to the East Coast by tanker from an ice-free port in southern Alaska and through the Panama Canal for about one dollar a barrel. Working back from selling prices for crude oil of around three dollars on the West Coast and four dollars on the East Coast, the profitability of North Slope oil is immediately apparent. From a production of two million barrels a day, the oil companies could expect to show a net profit of anywhere between $1–1.40 a barrel, while the state of Alaska would benefit in royalties and taxes by around $280 million a year, apart from additional amounts gained by selling oil leases. (The entire state budget for 1969 was little more than $150 million.)

This is what the oil boom meant, both to the oil industry and the state of Alaska. With the incentive of such enormous profits ahead of them, and in view of the fact that a large acreage of land around Prudhoe Bay was still available for leasing and due to be auctioned to the highest bidders at the next lease sale scheduled for September 1969, those companies without an existing stake in the area made every effort regardless of cost to test out the best locations. Geophysical teams swarmed all over the North Slope to carry out seismic surveys, wildcat wells sprouted on leases that some companies were fortunate enough to hold already. Meanwhile, BP, Atlantic Richfield and Humble Oil, comfortable in the knowledge that they owned most of the field through their existing leases, concentrated on solving the problem of getting the oil out. Whichever way it was to be, either by sea or by pipeline, would prove a tremendous undertaking. But the oil industry had never wanted for ingenuity in tackling such problems. Because something had never been done before served only as a challenge. And so, since a tanker route from the North Slope would be the most economic means of transportation if it was practical, the three companies concerned set up a joint experimental project to attempt what so many before them had tried to do but failed—the opening of the North-west Passage for commercial, year-round shipping.

12

North-west Passage

To the European merchant adventurers of the fifteenth century, lured by the riches of the Orient with its 'barbaric pearl and gold', the most eagerly sought prize was a short trading route between the Atlantic and the Pacific which would avoid the hazards of the long south-west voyage round Cape Horn and the even longer route eastwards. Columbus thought he had found it, but discovered America instead. Within a few years, mariners and explorers were trying to find a route round the top of the new continent to reach their original goal by what came to be known as the North-west Passage. Foi over 400 years they tried and failed, braving the storms and crushing ice-packs of the Arctic Ocean with an almost perverse tenacity in vessels that were sometimes no bigger than a modern cabin cruiser. The waters of the North Altantic were perilous enough, but at least the dangers which faced the seamen there were understood and accepted. The strange world in which they found themselves on sailing further northward was completely unknown, a labyrinth of narrow channels winding between islands, of inland seas and towering white cliffs, and, above all, ice—vast carpets of it heaved upwards into ridges and fantastically sculptured shapes, so that it was often difficult to tell where the sea ended and land began. But they were sure that such a passage existed. One channel after another was tried, only to end in dangerous cul-de-sacs from which many ice-trapped ships could not escape. The quest took a relentless toll of human lives and ships, but as each expedition went a little further than the last, they gradually charted and defined the Arctic regions. Although they did not succeed in their main endavour, the explorers opened up the northern areas of America and Canada for the traders and settlers who were to follow them.

The first mariners to make the attempt were mostly Italian and Portuguese, beginning with John Cabot in 1497, only five years after Columbus discovered America. Like Columbus, Cabot was also born in Genoa, Italy. His first voyage from England in the tiny ship *Matthew* took him as far as Nova Scotia. A year later he sailed even further northwards, but then he and his ship were lost off Newfoundland. The first men to attempt the North-west Passage were also its first victims. The challenge was taken up by Cabot's son, Sebastian, and others such as Jacques Cartier and Giovanni de Verrazano. But it was the British who became the most determined to find the new route, when for a period of nearly 300 years between the reigns of Queen Elizabeth and Queen Victoria it became a matter of national pride to be first to cross the Arctic Ocean.

Martin Frobisher was first, the privateer turned explorer, who in 1576 set sail from England and in two small ships of twenty tons each, the *Gabriel* and the *Michael*, and a pinnace of less than ten tons. He was obsessed by the idea of finding a North-west Passage as the only voyage of exploration 'left yet undone'. Although he got no further than what is now Baffin Island, he assumed from the features of the Eskimos he met that he had reached the Asian continent. As a result of some ore samples he brought back which were thought to contain gold, the Cathay Company was formed and two years later sponsored an expedition of fifteen ships under Frobisher, including men to mine the ore and settlers for the new land. Most of the ships were lost in storms, and by the time the few survivors straggled back the ore had been finally identified as worthless iron pyrite. The Cathay Company went bankrupt and that was the end of Frobisher's voyages into northern waters. He was discredited, in spite of the valid discoveries he had made, but eventually found the fame he had sought with such determination as one of the heroic band of seamen who opposed the Spanish Armada.

Even the Cathay fiasco could not quell the restless urge for adventure and profit that was so much a part of the Elizabethan character, and it was only a few years later that another explorer came along with the same quest in mind. John Davis was both a brilliant navigator and a scholar. During his voyages between 1585 and 1587 he went further northwards than anyone in recorded history, discovering the existence of a large inland sea to the west of Greenland that was later credited to another as Baffin Bay, and also noting the entrance to an even more important waterway that came to be known as Hudson Bay. The fight

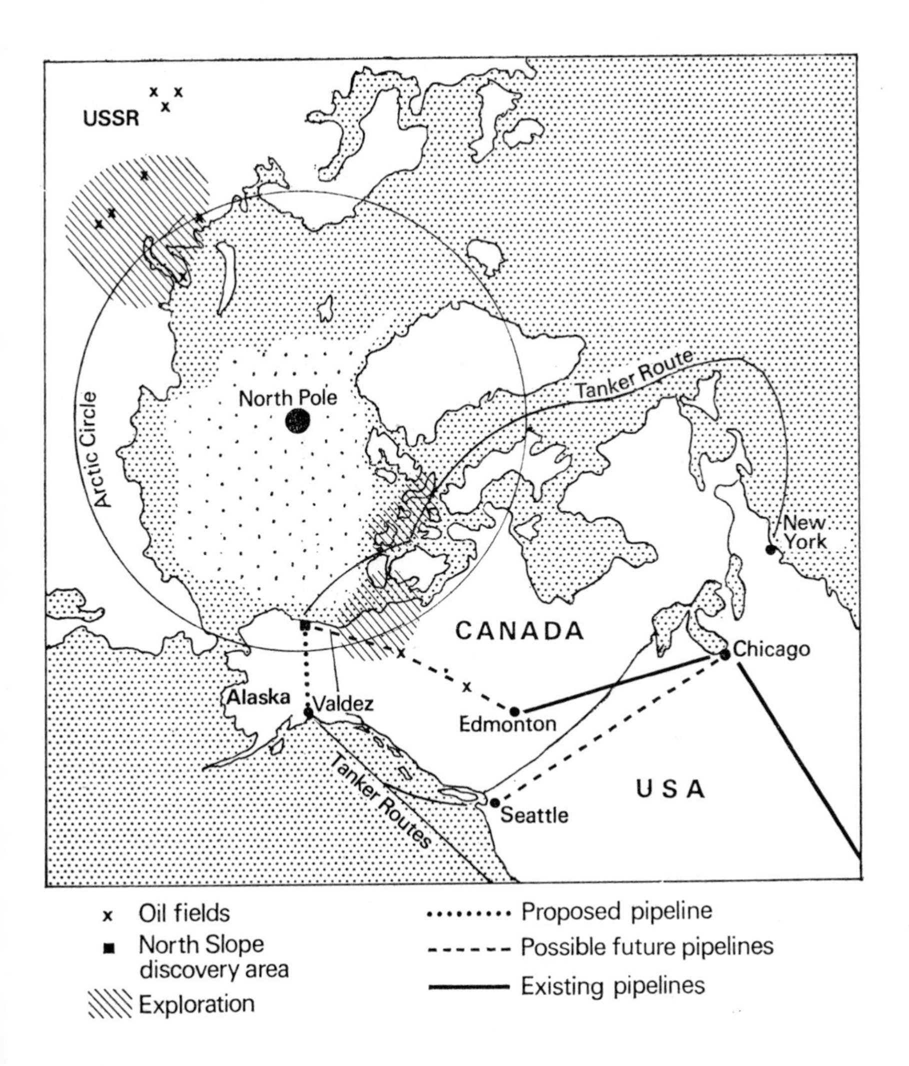
USSR
North Pole
Arctic Circle
Tanker Route
New York
CANADA
Chicago
Alaska
Valdez
Edmonton
USA
Seattle
Tanker Routes
Oil fields
North Slope discovery area
Exploration
Proposed pipeline
Possible future pipelines
Existing pipelines

The oil routes

against the Armada put paid to further explorations by John Davis, and it was not until after the turn of the century that the search was taken up again. This time it was a ship that achieved a fame equal to that of the explorers she carried. The *Discovery* was only fifty-five tons but best known of all the early ships that sailed in Arctic waters.

The first voyage northwards of the *Discovery* was in 1602 when George Weymouth was commissioned by the East India Company to explore the unnamed strait that Davis had noticed, and which appeared to offer the best chance of a passageway to the east. Weymouth had nothing but trouble from a mutinous crew, awed and frightened by the unknown icy waters of the Canadian archipelago, and he was forced to turn back after sailing a short distance through the strait. It was left to Henry Hudson to complete the voyage eight years later.

Hudson's tempestuous career as an Arctic explorer lasted for only four years, from 1607 to 1611. During that time he was the first to explore inland of what is now New York, sailing far up the river that bears his name. On a later voyage into Arctic waters he sailed to within 600 miles of the North Pole. And after entering the great inland Bay which was also named after him, his ship the *Discovery*, given to him after Weymouth's retirement, was the first to spend the winter in northern Canada after becoming stuck fast in the ice. He accomplished in a short time as much as any explorer before or after him. But whether because he was a poor judge or a poor leader of men, he continually had problems of mutiny from his crews. The culmination came on June 22, 1611, when the *Discovery* at last became freed of the ice that had held her trapped all winter. After months of cold and living on short rations, the crew expected to sail for home. Hudson was equally determined to sail across the bay to complete what he mistakenly thought was the North-west Passage. Most of the crew mutinied. Hudson, his young son John, five sick men and two others who refused to join the mutiny, were set adrift in a small open boat with neither provisions nor spare clothing. In such conditions they could not have survived for long. They were never seen again, and the manner of their death will never be known. Only a few of the mutineers survived the voyage home,the others either died from scurvy or were killed in a fight with Eskimos.

The next explorers to take the *Discovery* into Arctic waters were Robert Bylot and his navigator William Baffin. They charted amongst other discoveries Lancaster Sound, which ultimately proved to be the correct entrance to the North-west Passage. But it was still generally thought that the path lay through Hudson Bay, and it was there that later expeditions concentrated their efforts. Years of continued failure and frustration followed, until it was eventually proved beyond any doubt by ships charting the Bay that there was no practical outlet westwards. By that time, the opening up of Arctic waters by the early pioneers had led to other advantages. Britain became more concerned with exploiting the wealth of Canada's fur trade through the Hudson's Bay Company, and with developing her new possessions on the American continent, than in finding the elusive route to the Orient. The search was neglected for more than a hundred years. Only James Cook in the eighteenth century made an attempt to find the passage by approaching from the other side, sailing through the Bering Strait and eastwards along the coast of Alaska. He got as far as Point Barrow, but then met an impenetrable barrier of ice and had to turn round. It was not until after the end of the Napoleonic wars early in the nineteenth century that the quest was taken up again, this time officially by the Admiralty and ships of the Royal Navy.

The years between 1818 and 1859 saw the peak of Arctic exploration by such men as James Ross, William Parry, Frederick Beechey and John Franklin. Parry sailed 600 miles through the ice of Lancaster Sound, more than halfway through the passage. Ross and his men spent a remarkable four years in the Canadian Arctic, making overland explorations which resulted in the discovery of King William Island and the planting of the British flag at the magnetic North Pole. And Franklin in the 1820s made two notable land journeys to map the coastlines of Canada and Alaska, being the first non-native to set foot on the North Slope. He named many of the islands, bays, rivers and points of land along the north Alaskan coast by unmistakably British names that still remain today—even Prudhoe Bay itself. But he left his Union Jack flags at home, respecting Russia's ownership of Alaska.

These expeditions were carried out with all the scientific fervour of the nineteenth century, which led to the charting of much of the Canadian archipelago to the east. There were still unknown islands to the west but if a channel through these could be navigated to the Beaufort Sea beyond, it was known that a sea route existed between the Alaskan coast and the line of permanent polar ice. The ships were now larger, stronger and better suited to Arctic conditions, and the way at last seemed open finally to conquer the North-west Passage. For the great expedition of 1845 that was confidently expected to fulfil the dream of 300 years, Sir John Franklin was chosen as its leader. He was then fifty-nine years old.

Franklin set sail from England on May 19 that year in command of two ships of the Royal Navy, the *Erebus* and the *Terror*, with hand-picked crews of 168 officers and men. Both ships had emergency steam power. They carried provisions for at least four years and guns that could be used for hunting game. On July 26 they were sighted by a whaler heading towards Lancaster Sound. And that was the last ever seen of them. After three years had passed without any news from Franklin a relief expedition was sent out to search for him, comprising separate units under Sir James Ross which approached from three different directions: through Lancaster Sound on Franklin's track, from the Bering Strait to the east, and down the Mackenzie River and along the Canadian Arctic coast. They returned without finding a single clue. Over the next ten years the massive search was continued, urged on by Lady Franklin, who refused to believe that her husband had died. There were official expeditions organised by the Admiralty, others sponsored by Lady Franklin herself, by public subscription, and one by an American millionaire, Henry Grinnell. And as the result of a £20,000 reward offered by the Admiralty to anyone who could bring relief to Franklin, many private whalers and schooners joined in the search. All of them failed.

In 1854 an overland survey party, sent by the Hudson's Bay Company to examine the Canadian coast along the Boothia Peninsula, came across some Eskimos who showed them pieces of initialled silverware and other relics, including Franklin's star of the Order of Hanover. The Eskimos said they had obtained them from other natives who had reported seeing some 'forty white men travelling in company southward over the ice and dragging a boat and sledges with them' along the western shore of King William Island. These other Eskimos had apparently later discovered corpses and graves at the mouth of Back River, but it seemed that some of the white men had survived until the coming spring for shots had been heard and goose bones and feathers found. Dr John Rae, leader of the search party, purchased the relics and brought them and his report back to England. The Admiralty by now had amended its reward to £10,000 for information regarding Franklin's fate. It was no longer generally believed that he or any of his crew could still be alive. Rae had not previously known of the offer but he laid claim to it and eventually received £8,000, the rest being shared among his men.

Still the forlorn search continued, mainly at the desperate persistence of Lady Franklin, backed by a sympathetic public. Finally, in 1859, an expedition under the command of Captain Frederick McClintock discovered on King William Island a few gruesome relics, such as skeletons and uniform buttons, and a tattered piece of paper on which were scribbled two notes in different hands, the last dated April 25, 1848, and written by Captain James Fitzjames of the *Erebus*. It simply stated that the ships had been deserted on April 22, 'having been beset since September 12, 1846', and that 105 men under Captain Crozier had landed on the island. 'Sir John Franklin died on June 11, 1847, and the total loss by deaths in the Expedition has been to this date nine officers and fifteen men.' That was the only first-hand knowledge ever found as to the fate of the Franklin expedition. It remains one of the mysteries of the Arctic that so often has taken a terrible vengeance on those who dared trespass on her icy domain.

The tragedy cast a pall over Victorian England. With McClintock's return the search was officially ended, and so was a whole era of Arctic exploration. During the attempts to find Franklin a great deal more had been learned about the Arctic regions, filling in many of the missing pieces on the map. Robert McClure explored the western islands of the Canadian archipelago and discovered the Prince of Wales and McClure straits, the two routes leading into the Beaufort Sea which finally proved that a North-west Passage was possible. Rae discovered Prince Albert Sound; McClintock charted the last 120 miles of the north Canadian coast, thus completing the discovery of the coastline of continental America. The present-day map of the Canadian and American Arctic, dotted with the names of these men and of their sponsors and patrons, is a permanent memorial to one of the greatest periods of exploration. But having established the existence of the North-west Passage, Britain withdrew once and for all from the north. Never again would a British ship attempt the short cut from the Atlantic to the Pacific.

It was left to a young Norwegian, Roald Amundsen, to make the first voyage through the passage in 1903–6, slipping through almost unnoticed in the *Gjöa*, a former herring boat he had bought for the purpose. He had to leave Norway stealthily at night in order to escape a creditor who threatened to stop the expedition. In this small boat of only forty-seven tons, with a crew of six, he succeeded where hundreds of mightier ships before him had failed, just as later he was to beat Scott to the South Pole. His route took him through Davis Strait, west of Greenland, across Baffin Bay into Lancaster Sound, down through Peel Sound between Somerset and Prince of Wales islands, round the shallow waterways of King William Island, where he spent nineteen months making scientific observations, then through Dease Strait, Coronation Gulf and finally Dolphin and Union Strait into the Beaufort Sea. The point at which the strait widened between the Canadian coast and Banks Island was named Amundsen Gulf. He spent another winter in the ice off the mouth of the Mackenzie River and finally emerged through the Bering Strait and into the Pacific in August 1906.

The North-west Passage had finally been conquered after 509 years.

But there was no rush of ships to follow Amundsen. It seemed that the route would never be a commercial proposition. Long stretches of it were never entirely free of ice—only for a few short weeks in summer could the pack-ice of up to 100 feet thick be expected to break up sufficiently to allow vessels through, and even this was not a certainty if there was a late thaw or an early winter. The only ships that could plough their way through were specially reinforced ice-breakers with enormously powerful engines, developed by the Americans and Canadians for work in the Arctic. Several of these vessels made the voyage through the passage from 1940 onwards, as well as two US Navy nuclear-powered submarines, but no ordinary commercial vessels could do so. Or so it was thought, until the discovery of oil on the North Slope in 1968 gave a new impetus to the commercial possibilities of the North-west Passage.

Over the years a new type of vessel had been developed to meet the transportation needs of the modern oil industry. The oil tanker had grown from a 300-foot prototype launched in 1886 capable of carrying 3,000 tons of oil to 1,200-foot giants able to carry 300,000 tons, by far the largest craft ever built. In gross tonnage they represent over 36 per cent of the total world merchant fleet, although, in numbers, because of their great size, they account for less than 10 per cent of all merchant and passenger ships afloat. More than 50 per cent of the world's cargo being carried at any one time is oil, a graphic illustration of the world's dependence on this vital fuel. Tankers, many of them over 200,000 deadweight tons (their carrying capacity), are the only means of transporting crude oil in the quantities required from the producing countries to the consumer markets; pipelines can be used in those nations which have their own domestic sources of oil, such as the United States, but, even there, because of the greater flexibility of sea transportation as against a fixed pipeline which can only very expensively be altered once it is laid down, a considerable quantity of oil is shipped from the Texas Gulf to the East Coast. The reason for the rapid growth in the size of tankers—ten years ago 65,000 tons was top weight, now 500,000-tonners are being built and there are plans for craft of one million tons—is primarily one of economy. The bigger the vessel, the lower the relative cost per barrel of oil carried. Although a modern oil tanker of 200,000 deadweight tons might cost seven million pounds to build, her life will average about twenty years. During that time, for example, on a regular 5,000-mile route from a loading terminal to a refinery and back again in ballast, she would be capable of moving each year some two million tons of oil (fifteen million barrels, 630 million gallons).

Generally speaking, the cost of transportation by large modern tanker is considerably less than a pipeline operation. It is not often possible to compare the two from an economic point of view, for usually there is no question of choice in the method to be adopted; pipelines being used for moving oil inland, tankers for shipment from one country to another. But the North Slope was one of the exceptions. The oil discovered on the coast of Arctic Alaska is just about as far away as it could be from the American markets most in need of it—California to the west and New York to the east. Transportation was obviously going to be the biggest problem of all, but when, in 1968, the oil companies began seriously to study ways of moving the oil out they found that at least they had the choice of two possibilities, by pipeline or tanker. There were other suggestions. A rail link from the North Slope to Anchorage, for instance, to freight the oil down in tank wagons. Or submarine freighters which could navigate under the Arctic ice, proposed in a detailed study by General Dynamics. But in terms of the amount of oil to be moved—at least two million barrels a day, and possibly more in future—these were just not practical After analysing the costs involved, it became apparent that shipment by tanker to the major markets would be as much as half the cost of a pipeline, a saving of up to sixty cents a barrel, over a million dollars a day. It would require a fleet of between thirty and forty tankers of around 200,000 tons, specially constructed to withstand Arctic conditions and costing about two billion dollars. It would also mean achieving what mariners for 500 years had tried to do and failed—opening up the North-west Passage as a commercial shipping route.

In mid-December 1968 the three companies most closely concerned, Humble Oil, Atlantic Richfield and British Petroleum, announced a dramatic experiment. They would try to send a tanker from Chester, Pennsylvania, to Prudhoe Bay to test the feasibility of a route through the North-west Passage. Humble was to put up most of the forty million dollars required for this, Altantic Richfield and BP contributing two million dollars each. The marked difference in the sums of money put up for the experiment by the American partners in the North Slope discovery and BP reflected their differing degrees of optimism. From the very start, the British were dubious about the project. It was not so much a matter of whether tankers could physically get through on the regular three-month round trips projected by Humble, although unconsciously perhaps there may have been an awareness of all the failures in the past to conquer the passage. The biggest problem was how to load the oil on to tankers. A normal jetty and loading terminal was out of the question. Because of the shallow waters off Prudhoe Bay, tankers of the size envisaged could not anchor closer than twenty-five miles from the shore. A pipeline would have to be laid on the sea-bed out to a fixed or floating platform, but this would be subject to tremendous pressures from pack-ice as it built up in winter and shifted in storms and currents. It was more likely that tankers would moor in the Mackenzie River delta to receive oil from a pipeline laid along the coast, but even the thinner ice there could cause similar problems. There was also the question of Canadian sovereignty over the North-west Passage, whether or not it lay in international water or Canadian territorial water, since the route through the more protected Prince of Wales Strait was less than six miles wide. It had never been an issue before, but now questions began to be raised in the Canadian Parliament. Many naturalists were worried about the consequences of an accidental oil-spill, and after the *Torrey Canyon* disaster in England this became a matter of widespread public concern. The effect of such pollution in the icy waters of the Arctic, where the oil might not disperse for many years, could be catastrophic. The Canadians, if it is accepted that they have territorial rights over even part of the route, might insist on such traffic and pollution controls that could add considerably to the expense of a voyage.

From the American viewpoint, on the other hand, a tanker route through the North-west Passage would have a major additional advantage. Because of high costs at home and competition from abroad, particularly Japan, the American shipbuilding industry has been in a state of decline since the end of the Second World War and the US maritime fleet has fallen drastically in comparison with those of other nations. The industry has been supported from even further economic eclipse by the protectionist Jones Act of 1920 which prohibits the movement of cargo between US ports on foreign-built or foreign-flag ships. Most of the ships built in American yards are for such coastal trade, generally small cargo vessels and barges. Since most crude-oil movements within the United States are by pipeline, there has been no demand for the large tankers that other maritime nations have been building. But the requirement that any tankers carrying oil from the North Slope to the East and West Coasts had to be US-built would mean an unprecedented boom for the American shipbuilding industry. The higher costs involved would not, of course, be to the favour of the American oil industry, whose lobby has on a number of occasions battled with the maritime lobby over the Jones Act, in particular with regard to the existing shipment of oil from the Cook Inlet region of Alaska to the West Coast. Alaskans themselves, their fishing and ferryboat operations hampered by the restrictions governing foreign-built vessels, have for years been trying to have the Jones Act repealed.

However, regardless of all the problems, it was first necessary to test whether a tanker could make the voyage through the North-west Passage. For the Arctic Marine Expedition, as the project was called, Humble chose the largest tanker in the US merchant fleet, the 115,000-ton SS *Manhattan*. Built in 1962 by the Bethlehem Steel Company at their Quincy, Massachusetts, yard for Stavros Niarchos as part of a complicated deal by which the Greek shipowner was then entitled to transfer six smaller American-flag tankers to foreign registry, the *Manhattan* had from the outset been something of a white elephant. Bigger than any other tankers being built at that time, with a length of 940 feet and a draft of over fifty feet, she was unable to dock at most American ports. Because the design of such a large-size vessel was new, more steel was used in her building than was later found to be necessary; she was stronger, in fact, than even larger tankers built afterwards. And her 43,000 shaft horsepower steam-turbine engines, driving two propellers for greater manœuvrability, were also more powerful than those of most other tankers, giving her a speed of 17·5 knots. They were not factors conducive to an economic operation. After costing twenty-eight million dollars to build, Niarchos sold her within a year to Seatrain Lines Inc., who then used her mostly for carrying grain to India. But these same factors of great strength, manœuvrability and speed made her ideal for the purpose now intended. Humble chartered the *Manhattan* from Seatrain—and promptly began to redesign her.

The modifications carried out were on such a scale that only an industry accustomed to thinking big could have attempted them. They involved strengthening the ship internally with a web of steel braces, building a second hull round the potentially vulnerable engine room, fitting new propellers and ice deflectors to protect the rudder, welding a sixteen-foot high eight-foot thick belt of steel round the outside of the hull to alleviate the pressure of tons of ice crushing against the sides and to prevent the ice from climbing up on to the deck, increasing the accommodation space to take 126 instead of the normal sixty crew, erecting laboratories and a helicopter pad on the deck, and replacing the bow with an entirely new section specially built on traditional ice-breaking principles to enable the prow to ride up on the ice and then crush it with the weight of the ship. This perhaps was the most vital addition of all. Designed by the Massachusetts Institute of Technology in co-operation with the US Coast Guard, the new prow curved up from the keel at an angle of 30 degrees and had massive shoulders of steel jutting out eight feet from either side of the hull to open a wide path for the rest of the ship.

In the short time available—there were only seven months before the voyage was due to begin in July 1969—no single shipyard was able to undertake such a mammoth task. And so the tanker was literally cut into four sections at the Chester, Pennsylvania, shipyard of the main contractor, the Sun Shipyard and Drydock Company. The bulkheads were sealed and the forward and midship sections were towed to other yards in Virginia and Alabama for conversion. The major stern section remained at Chester, and it was there that the parts were later fitted together again, including the new prow. The *Manhattan* was a different ship, the strongest and most strangely shaped tanker ever built. Her length had increased to 1,005 feet, her beam from 132 feet to 155 feet, and her draft from just over fifty feet to fifty-two feet. In one respect she was smaller: because of the addition of protective steel her deadweight tonnage, being the amount of cargo she could carry, was reduced from 115,000 tons to 105,000 tons. But the extra weight of steel had increased her displacement tonnage from 137,068 tons to nearly 150,000 tons. This was the ship that would make the first commercial attempt to sail through the North-west Passage. And since, if the venture was to be a success, later ships would have to make the voyage in all weather conditions and at all times of the year on a rigid schedule, it was no use the *Manhattan* trying to skirt round the most difficult areas as other ships had done. She would have to meet the ice head on and overcome it if the route was to be proved a practical proposition.

The *Manhattan* set sail from Chester on the start of her 4,500-mile voyage shortly after eleven o'clock on Sunday morning, August 24, 1969—more than a month late—with a crew of fifty-four and a further seventy-two scientists, Canadian and American government representatives, oil-company officials, and news and cameramen, for, whatever happened, this was going to be the most publicised attempt on the passage ever made. In command was Captain Roger Steward with two staff captains, Donald Graham and Arthur Smith, and Stanley Haas, the Humble executive in charge of the expedition. The two BP observers on board were Captain Ralph Maybourn, the company's chief marine superintendent, and Paul Heywood, a naval architect. Among the scientists was Dr Elbert Rice of the University of Alaska, whose team were to carry out measurements and other tests on the ice. The route that had been selected would take the tanker up the eastern coast of the United States to Nova Scotia, with a stop at Halifax, then across the Labrador Sea and along the coast of Greenland, turning westwards at Thule to cross Baffin Bay and entering the 1,200-mile North-west Passage through Lancaster Sound. After following the Parry Channel to Viscount Melville Sound there were two possible exits into the

Beaufort Sea, southward through the Prince of Wales Strait or north-westward through McClure Strait, which had never yet been navigated by any ship. The decision on which route to take would depend on the conditions when they got there.

The first of the icebergs that infest the Labrador Sea and the coastal waters of Greenland were sighted early on the morning of August 30. From then on, ice in one form or another was always present. The icebergs, awe-inspiring though they were, with their towering peaks and the knowledge that the greater part of them lay dangerously hidden beneath the surface of the water, could be avoided with reasonable ease by sight and radar. But then came the first of the pack-ice, off Baffin Island, which instead of being avoided as would normally be the case had to be rammed to see how the ship would behave. By this time, early in September, the Canadian ice-breaker *John A. Macdonald* had joined the expedition and would remain with the *Manhattan* for the rest of the voyage. While she stood by to one side, the big tanker gingerly nosed her way into the ice at about two knots. The massive prow crushed the floes and heaved them apart with almost contemptuous ease. When the speed was increased to twelve knots, boulders of ice were tossed aside with barely a hesitation. The first test of *Manhattan*'s ice-breaking abilities was completely successful. But this was relatively thin and unprotected ice in open sea, very different from the thick, concentrated polar pack-ice of the actual passage itself. Ice-floes, roughly circular in shape, could be miles in diameter, compressed by winds and currents to form pressure ridges of ten or more feet high and as much as 100 feet below the surface of the sea.

When the first of these massive ice-floes was encountered in the Viscount Melville Sound during the second week of September the *Manhattan* became stuck and had to be broken free by the *Macdonald*. This became the pattern for much of the remainder of the voyage through the passage. The *Manhattan* was able to plough on steadily through ice of up to five feet, but the thicker, older floes would bring her to a stop. The technique was to reverse and charge repeatedly in an attempt to break through. It was then that the advantage of weight, thought to be the greatest asset of giant tankers in combating ice, became a liability, for without forward momentum the *Manhattan* was ponderous and sluggish. With only 17,000 horsepower available for going astern this was often not sufficient to drag her free from the embedding ice. The *Macdonald* would have to approach from behind and cut into the ice alongside to release the pressure. This became a well-rehearsed routine, especially when the decision was taken to try the much harder route through the 200-mile McClure Strait than the relatively ice-free Prince of Wales Strait. No previous ship had ever navigated the McClure Strait and it was now completely covered with ice, much of it old and tough polar floes. But the voyage was meant to be an experiment and it was necessary to see how the *Manhattan* could cope with these most difficult conditions of all, which at other times of the year could be expected throughout the passage.

Constantly requiring the assistance of the much smaller ice-breaker, the tanker fought her way nearly halfway through the strait. But when the *Macdonald* herself became seriously endangered it was necessary to turn back. This in itself was one of the most difficult and dangerous operations of the voyage, in which the tanker had to back and charge forwards time and time again to extricate herself from the powerful grip of the ice, while the *Macdonald* did the same, to break a path through. The attempt to force a way through the McClure Strait had taken three days, and in that time a change of wind had blown huge ice-floes into the narrow Prince of Wales Strait that would be the likely route of any regular voyages in future. There were further hazards as the two vessels edged through, sometimes at no more than a mile an hour. But on the afternoon of September 14 they finally broke through the pack-ice into open summer-blue water. Three days later, after a short stop at Sachs Harbour on the southern coast of Banks Island, one of the few entirely self-sufficient Eskimo settlements in the Arctic, the *Manhattan* entered Alaskan waters. After anchoring off Prudhoe Bay for the ceremonial loading by helicopter of a gold-painted oil-drum the *Manhattan* reached Point Barrow, the western end of her voyage, on September 21. The original plans had called for a symbolic completion of the voyage through the Bering Strait to Anchorage, thus connecting the Atlantic and Pacific oceans, but because of the six-week delay in setting out this had to be cancelled. The tanker turned about and headed back by the same route, eventually making a triumphal entry into New York harbour on November 12.

In the sense that the *Manhattan* had been the first cargo-carrying vessel to conquer the North-west Passage, the voyage was a historic occasion. There were hopes not only that an oil route could be established but that the other vast resources of the Arctic, such as iron, copper, nickel, silver and uranium, could be developed, creating cities and settlements in the barren northland. But when the great amount of data collected on the voyage was analysed, doubts began to creep in. For one thing it was apparent that the *Manhattan* could not have made the voyage alone without the help of the *John A. Macdonald* and the other ice-breakers that joined her at various times. Admittedly, future tankers would be bigger and better constructed for Actic work. But how would even they fare in the severe conditions of winter? The *Manhattan*'s voyage was difficult enough, and that had been undertaken at the most ideal time of the year, towards the end of the short Arctic summer when the passage was as ice-free as it could ever be. Even a few days later, at the start of the voyage back, the *Manhattan* encountered ice-floes in what had previously been open water and the sea was already beginning to freeze over. And then there was the even bigger problem of how to build a loading terminal that had been apparent from the outset. Further tests and engineering studies were carried out, but with a growing realisation that for the foreseeable future at any rate the North-west Passage was not a commercial proposition. During the summer of 1970 the *Manhattan* carried out some further trials in Arctic waters. And then, her moment of glory over, she was quietly returned to her owners to resume her more mundane duties as a grain and occasional oil carrier. She may have made history on two counts, not only being the first commercial ship to sail the North-west Passage but also the last, proving after more than 500 years the final victory of the Arctic ice over man.

Meanwhile, the oil companies had turned their main attention towards the alternative method of moving North Slope oil by pipeline. As early as 1964 they carried out a preliminary study which seemed to show that the project of building a pipeline across Alaska to an ice-free port in the south was feasible, although it also anticipated considerable problems such as the presence of permafrost over most of the 800-mile route, severe climatic conditions, with temperatures varying between minus 70°F in winter to plus 90°F or above in summer, and the logistical difficulties or transporting and maintaining men and equipment in a bleak and inhospitable terrain.

Little more was done at that time; it was sufficient to know that there was a possible way of getting oil out if a discovery was made. When this proved to be the case in 1968 with the Prudhoe Bay finds, pipeline experts from BP, Humble and Atlantic Richfield got together to examine the project more closely. With Pipe Line Technologists Inc. called in as consultants, they began map studies and aerial reconnaissance over a number of possible routes, while other specialists investigated potential marine terminal sites along the western and south central coasts of Alaska. These examinations demonstrated that Prince William Sound, and in particular the harbour at Valdez, offered the necessary requirements of deep, sheltered and ice-free tidewater with enough width and length for manœuvring large ships, and that it was also accessible by a pipeline of the shortest length from Prudhoe Bay. In August 1968 field parties were sent out to survey the general route of the proposed pipeline, which fell into four main areas: Prudhoe Bay to Anaktuvuk Pass in the Brooks Range, Anaktuvuk to the Yukon River, Yukon to Fairbanks, and Fairbanks to Valdez through the Copper River Valley. An alternative route was examined southwards from Prudhoe Bay along the Sagavanirktok River, over a high unnamed pass across the Brooks Range, and down into the Dietrich River Valley to join the main route south of Bettles. This would present more severe hydraulic problems, since the pass had a maximum altitude of 4,700 feet compared with the 2,400 feet of Anaktuvuk Pass, but avoided much of the North Slope permafrost by following the gravel deposits of the Sagavanirktok flood plain.

In February 1969 the companies announced their plans for the forty-eight-inch trans-Alaska pipeline, a larger-diameter pipeline than had ever been built before, in which BP and Altantic Richfield would each have a 37½ per cent interest and Humble Oil 25 per cent. The estimated cost of the 800-mile line would be about $900 million, with completion scheduled for 1972 at an initial capacity of 500,000 barrels a day. So confident were the companies that federal and state government permission to build the line would be given that world-wide bids were invited for the supply of the steel pipe required, and in June contracts were awarded to three Japanese companies: Sumitomo Metal Industries, Yawata Iron & Steel, and Nippon Kokan. Only in Japan could the size of pipe be made in the quantity required and in the time available. As an interesting sidelight, it was apparent that even before the voyage of the *Manhattan* began the companies had plumped in favour of a pipeline, and the great North-west Passage project had developed into something of a public relations exercise to show the government and the maritime lobby that at least it had been tried. Even if successful it would have taken years to build up the necessary tanker fleet, and the companies were eager to start moving North Slope oil as soon as they possibly could.

By September the main surveys had been completed and the engineering staff based in Houston, with site offices in Anchorage and Fairbanks, got down to detailed design work. Five other companies decided to take a share in the pipeline—Mobil (8·5 per cent), Phillips and Union of California (3·25 per cent each), Amerada-Hess (3 per cent) and Home Oil (2 per cent); to accommodate them, BP and Atlantic both agreed to reduce their shareholding by 10 per cent to 27½ per cent each. Home Oil later withdrew from the project. The first supplies of steel pipe began arriving from Japan, and with Alaskan industry poised to take its biggest share yet in the oil boom by helping to build the pipeline and the ancillary road alongside that would be necessary to move in men and equipment, everything seemed to be going with the speed and efficiency to which the oil industry was accustomed. The protests of a few conservationists and native leaders were lost in the screaming headlines of the oil boom where everything was excitingly bigger than life. The discovery itself, the biggest-ever in America, with its dramatic stories of men battling the Arctic wilderness; the romance of the *Manhattan*'s voyage through the North-west Passage; the trans-Alaska pipeline which would be the greatest single engineering feat ever undertaken by private industry. No one dreamed that such a Colossus could be halted dead in its tracks by such seemingly unrelated considerations as wildlife preservation and the land claims of a few thousand Eskimos and Indians. September 1969 was the peak months for the oil industry's great endeavour, when everything was going right. And the biggest event was yet to come with the grand lease sale to which a year of almost superhuman effort had been geared.

13
The Big Sale

September 10, 1969, was the day that Alaska got rich. It was the day of the big lease sale, one of the strangest of American business rituals when, in an atmosphere part carnival and part shrouded in mystery, land that is thought to contain oil is auctioned to the highest bidders. In this particular case it turned out to be the biggest auction of any kind in history.

For days before the sale oilmen from all over the world had been gathering in Anchorage, crowding the hotel lobbies and motels to overflowing. With the tourist season over it was a welcome boom in itself to the city's hoteliers and bar-keepers. Big raw-boned men from Texas rubbed shoulders with nattily suited businessmen from Japan and smooth-talking executives from New York and London, all seeking a piece of the action in what had become the hottest oil play in the world. Groups of company men—directors, accountants and geologists—huddled together in corners to plan their strategy, those already operating on the North Slope jealously guarding the geological information they had obtained, their manner infuriatingly sure and superior to the 'have-nots' who could only guess at what were the best locations and what they might have to bid to get them. Rumour and counter-rumour ran riot as industrial spies tried to break through the tight cordon of security around the North Slope drilling operations.

On offer were 451,000 acres of previously unleased land around Prudhoe Bay and the Colville River. It was the fourth time that the state had put up North Slope land for lease sale since selecting in 1964

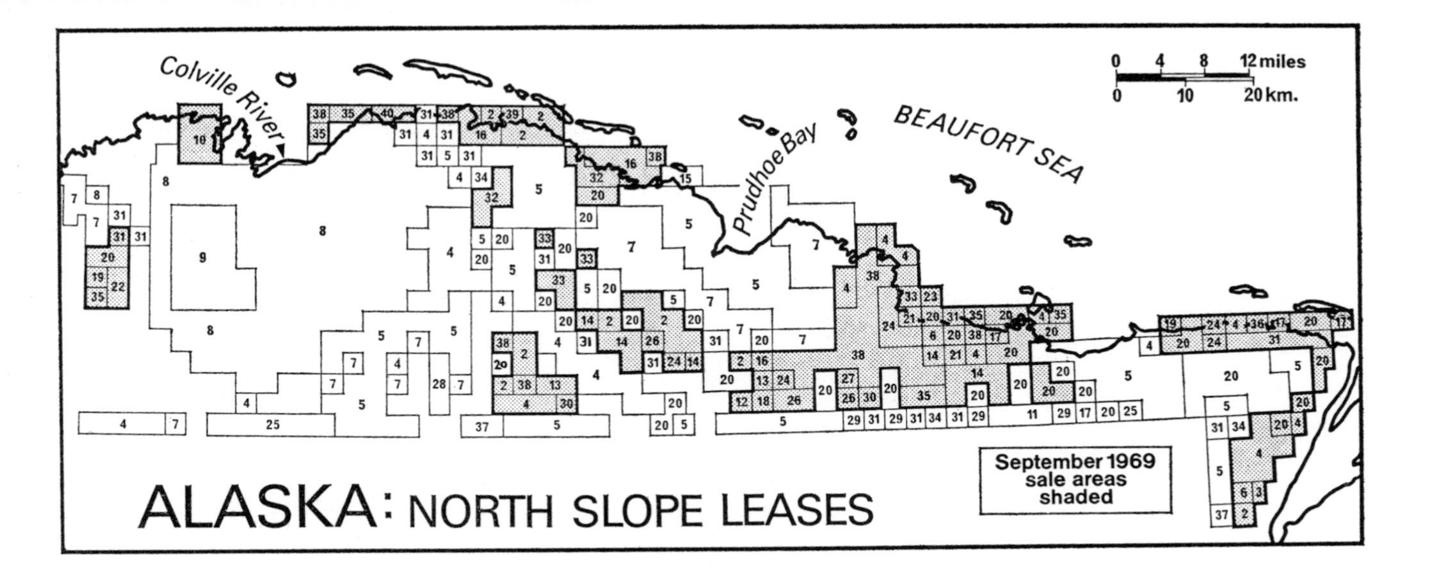

NORTH SLOPE: PRINCIPAL LEASEHOLDERS

Code		Code		Code		Code	
1	Alexco	11	Burglin	21	Nova Petroleum	31	Socal
2	Amerada-Hess *et al**	12	Burnett	22	Occidental/Agip/Buttes	32	Socal/Mobil
3	Apache	13	Champion	23	Panocean *et al**	33	Socal/Phillips/Mobil
4	Arco	14	Conoco/Sun/Cities Service	24	Pennzoil *et al**	34	Texaco
5	Arco/Humble	15	General American	25	Phillips	35	Texaco/Shell
6	Aztec	16	Hamilton Bros. *et al**	26	Priest	36	Tipperary Land
7	BP	17	Humble	27	Ranger	37	Union Oil
8	BP/Arco	18	Marshall	28	Rogge	38	Union Oil/Pan Am
9	BP/Arco/Union Oil	19	Mobil	29	Shell	39	Union Pacific/Signal *et al**
10	BP/Gulf	20	Mobil/Phillips	30	Shottmeyer Bros.	40	Waco group

*Participants in the indicated groups are usually as follows:
With *Hamilton Brothers*—Union Pacific, 1409 Corporation, Home Oil, Conoco, Sun and Cities Service.
With *Panocean*—PetroLewis, R. Duncan and V. Duncan.
With *Pennzoil United*—Forest, Colorado Oil & Gas, Newmont and Aquitaine.
With *Union Pacific/Signal*—1409 Corp, Union Carbide, Highland Resources, Conoco, Sun and Cities Service.
Participating with *Amerada-Hess* (in a variety of combinations) are Louisiana Land & Exploration, Marathon, Getty, Placid and Hunt.

two million acres of former federal land along the 140-mile stretch of coast between the Naval Petroleum Reserve and the Wildlife Range. Within that area, although a total of thirty-seven wells had been drilled from twenty-three active rigs and thousands of square miles of the whole of the region had been mapped by twenty-four different seismic crews, no company knew what others might bid for the leases offered, nor what it would have to bid itself to be sure of obtaining areas it particularly wanted. A careless word could cost a company valuable leases; on the other hand, a carefully planted rumour could influence a competitor to waste his resources by over-bidding for possibly worthless leases, leaving others free to be picked up more cheaply. Each company had only a certain amount of money it could spend. But there was no doubt, in view of what had been discovered already on the North Slope, that the bidding was going to be very high. The total twenty-two lease sales which had been held in Alaska during the state's ten-year history, including the previous three on the North Slope but the others mainly in the Cook Inlet area, had netted a total of $97·6 million. There was speculation that this twenty-third sale might alone realise the magic figure of one billion dollars, beating the previous record in the United States of $603 million, paid the previous year for 363,000 acres off the Californian coast in the Santa Barbara Channel. Some Alaskan officials tended to play the sale down, talking in terms of an expected $100 million. Even that would be a welcome addition to the state's treasury, almost as much as the entire revenue for the year apart from federal grants.

At seven o'clock on the morning of September 10 representatives of the various oil companies, together with individuals who were also entitled to bid for any of the 179 leases on offer, filed into the Sydney Lawrence Auditorium, the 350-seat hall named after the doyen of Alaska's artists in which the sale was to be held. In the presence of armed guards and state troopers they handed over the sealed envelopes containing their written bids, each envelope marked with the number of the tract it concerned, most of which consisted of 2,560 acres. From that moment on it was out of the bidders' hands. A lease would go to the company that had bid the highest, unless the state authorities decided afterwards that none of the bids on certain leases were big enough, in which case they would all be rejected and the leases withdrawn. There could be no raising of bids. All that remained for the state officials, sitting at a long table across the platform, was to sort them into the numerical order of the tracts offered and then to open the envelopes one by one when the sale started. A large board behind the platform would identify each tract as it came up and keep a tally as the bids were revealed. By the time the ceremony started, at eight o'clock sharp, with the singing of the national anthem and a short address by the state's governor Keith Miller, the packed audience of bidders and newspapermen and curious onlookers was in an almost festive mood, waiting with eager anticipation for the results. Some companies who were not personally represented had set up direct phone lines to the auditorium where a monitor would relay a bid-by-bid account to their headquarters in other parts of the United States.

It had been ten years almost exactly to the day that Alaska's legislature had passed regulations for oil and gas leasing on state lands that made the occasion possible. It was not surprisingly one of the new state's first actions since it was the Swanson River discovery, proved as a productive oilfield in 1959, that tipped the balance towards the granting of statehood that same year. Before then, companies and individuals had leased only from the federal government through the Bureau of Land Management. Now they would continue to deal with the Bureau over federal lands—which were still by far the greatest proportion of the whole region—but with the state authorities over state land. As in all other American states, Alaska automatically owned 'submerged lands' under rivers and lakes and tidal areas offshore as far as the outer Continental Shelf. It also had the choice of selecting from the maximum of 103·5 million acres available from federal land. One of the first areas selected was naturally on the Kenai Peninsula where oil had been discovered. The Cook Inlet belonged to the state anyway, as submerged land.

Another area selected was the coastal belt of the North Slope between the Naval Petroleum Reserve and the Wildlife Range, both federal reserves. Since the selection was made four years before any oil was found there, it might seem to have indicated remarkable foresight on the part of the state authorities. Some did in fact believe firmly in its oil potential—in particular Tom Marshall, the state petroleum geologist. But the real reason for the selection was much more practical and mundane. Because of the myriad lakes along the coast and the difficulty of measuring the tide of the Arctic Ocean which rose and fell only a few inches, it was a great problem to know where to draw the line between federal land and submerged land that belonged to the state. A few oil companies wanted to take out leases in that region, but were unable to do so until the matter was sorted out legally—and that might have involved years of delay in the courts. At the companies' request, therefore, the state agreed to take over the area as part of its selection, removing any cause for argument. As events proved it was a wise decision, but one criticised at the time as wasting the state's entitlement on a worthless piece of ice.

The state stood to gain a great deal by selecting lands from which oil was being produced or which might be produced at a later date. It could establish its own methods of leasing and retain all the income collected by way of bonus bids, land rents, and royalties and taxes due on any production. In the case of federal lands within Alaska it would receive only 90 per cent of the revenue, the other 10 per cent being retained by the federal government. Alaska is in a unique position among American states as a whole. With such a small population and with only a fraction of the land privately owned, there are basically only two landlords—the state government and the federal government. It does not have to contend with the political influences of private landowners on whose property oil is found and who want their own share of the industry's profits as in the other oil-producing states. In this sense Alaska bears a much greater resemblance to those developing countries of the world where oil has been found, such as in the Middle East and Africa where governments deal directly with the oil companies, than with the rest of the United States, a point strongly made by those who are critical of the way Alaska has handled its leasing laws.

There are several systems of oil and gas leasing in the United States. In the case of any federal or state land which is a 'known geologic structure' producing or known to contain petroleum, competitive bidding is usually employed. A state may also apply the system to land on which oil or gas has not actually been found but where it is thought likely to exist. The decision depends on the degree of interest being shown by the oil industry, it being the aim of the government, whether state or federal, to encourage private companies to invest in the costly gamble of exploration. The first step under competitive bidding is for one or more companies who have an interest in a particular area to request that acreage be put up for nomination. The authority concerned blocks out the tracts it decides to make available and publishes these on a map of the region. It is then the turn of any companies or individuals to consider what the leases might be worth to them, and to make an offer for them in the form of bonus bids. This is in addition to the lease rental which is payable anyway by the successful bidder. The bonus offers are made confidentially in sealed envelopes which are opened at random on the day appointed for the sale, when the apparent highest bidder for each lease or block of leases is declared.

Apart from the ordinary filing of leases on unreserved federal land on a first-come first-served basis, as described in Chapter 9, another system of leasing is by simultaneous filing. As with competitive bidding this is usually on land, either federal or state, which is known to be a petroleum prospect and therefore not open to the normal filing of leases. Acreage is nominated and put up for sale as before. But in this case there is no question of bonus payments. Applications sent in over a period of time from when notice was first given, normally thirty days, are considered to have been filed simultaneously. The leases are literally raffled, going to the first applications taken out of a hat. Only the ordinary rental is then payable, in addition, of course, to royalties if oil is later produced. The lottery of simultaneous filing is held to be a great democratic venture, giving a single individual or a small company as much chance of leasing land as a big corporation. In the case of an individual, he would probably seek to sell his leases to a company interested in exploring the area for a cash sum and an overriding royalty on any oil found.

The system of competitive bidding is usually the most profitable to the government. But it does not always attract the kind of record-breaking figures that make the headlines. These are the exception rather than the rule. The twenty million acres of the North Slope between the Naval Petroleum Reserve and the Wildlife Range, mostly geologically favourable for oil and gas formation and accumulation, were first opened to mineral leasing by the federal government in 1958. Interest naturally centred in the vicinity of the Umiat and Gubik discoveries that had been made by the US Navy during their exploration programme, and in the first competitive leasing in Alaska on September 3 of that year, bids of between $3 and $103 an acre were made for land near the Gubik gasfield. Other federal land was leased under simultaneous filing. But then the Swanson River discovery on the Kenai Peninsula directed attention away from the North Slope. Most of the American companies concentrated on exploration in the Cook Inlet region, in some cases reaching the limit there of the total acreage they were allowed to hold in Alaska. An exception was British Petroleum, which continued to lease federal land for its early drilling activities in partnership with Sinclair. And so also did a few farsighted individuals, such as Locke Jacobs in Anchorage and Cliff Burglin, the owner of an office-equipment store in Fairbanks. He acquired leases for as little as twenty-five cents an acre for three years' rent and turned an investment of a few thousand dollars into a fortune, like Jacobs also becoming an oil-lease broker.

The rules and regulations chosen by the state of Alaska insist that all tide and submerged lands, together with some other categories such as school and university lands, must be leased competitively. In addition whenever oil or gas is discovered, whether on state lands or not, the state Commissioner of Natural Resources determined the extent of the area around the discovery well which is likely to be productive and therefore classified as competitive. He can also offer lands on a competitive basis where it is in the 'best interest of the state'. The annual rental for a competitive lease of ten years is a dollar an acre. Upland leases may not exceed 2,560 acres and tide and submerged land leases are limited to a maximum of 5,660 acres. State lands which are not classified as competitive are non-competitive, whose leases have a five-year primary term, eligible for a two or five years' extension, with an annual rental of fifty cents an acre and a lease-filing fee of twenty dollars. In practice most of these come under the simultaneous filing system, but otherwise leases are available on a first-come first-served basis. The royalty on production is basically similar to that charged by the federal government—12½ per cent of the wellhead value of oil or gas. But in addition Alaska charges a severance tax, which is virtually a tax on production. This was originally set at 1 per cent and since increased to 4 per cent with current moves to raise it even further. On the other hand the state has provided some advantages over the federal system in order to encourage exploration by the oil industry. The royalty is reduced to only 5 per cent for the first ten years in the case of a first discovery in a new area, and the amount of land that may be held by one company is greater than in other states—a total of 500,000 acres onshore and a further 500,000 acres offshore, in addition to the maximum individual holding of 600,000 acres of federal land, including options.

Since first selecting its two million acres of land along the coast of the North Slope, the state had held three competitive lease sales there in 1964, 1965 and 1967 which brought in some twelve million dollars in bonuses for 906,842 acres, an average of little more than twelve dollars an acre. In some cases, bids as low as $1.55 an acre had been accepted. This had given individuals the chance of obtaining leases even under competitive bidding and a number of shrewd investors had done so, as well as holding leases filed on federal land before the freeze. One such individual was Tom Miklautsch, a forty-one-year-old Fairbanks pharmacist who in 1967 picked up two leases on the coast five miles from Prudhoe Bay for $4,800, plus an additional $4,800 a year payable in rent. He was one of an estimated 3,000 people in Fairbanks reckoned to hold oil-leases, out of a total population of under 40,000. After the second Atlantic Richfield discovery, prices began to move sharply upwards as new companies moving into the area sought options on leases from anyone who had them to offer. In the summer of 1968 prices rose from about ten dollars an acre to $100 an acre which Pan American had to pay for a lease on which it wanted to drill a well. In addition the overriding royalty that would be payable to a leaseholder in the event of petroleum being produced rose from 1 per cent to over 6 per cent.

By the end of the summer the oil companies had obtained all the land that was then available in the Prudhoe area except for the two leases held by Miklautsch. He continued to hold out as the offers became more and more tempting—including a certified cheque for one million dollars that was actually laid before him by one company. A Canadian oil group offered six million dollars in cash. But Miklautsch and his oil-broker and friend, Cliff Burglin, were more interested in teaming up with a company that would share the profits from developing the lease. Eventually, after lengthy negotiations, they came to an agreement in January 1969 with the General American Oil Company providing for a down payment of two million dollars in the company's stock, a larger but unspecified payment if and when oil was produced, and a 25 per cent share in all future production profits, from which Burglin would take a 20 per cent commission. It was the biggest deal of its kind involving an individual leaseholder that anyone in the industry could remember.

With the knowledge that the Prudhoe Bay oilfield was known to contain between five and ten billion barrels of oil, the companies got down to the task of assessing what they would have to bid for the remaining leases there at the September 10 sale. Costs were the all important factor. Because of its remoteness and severe climatic conditions these were much higher on the North Slope than elsewhere in the United States, one company estimating drilling averages at $142 a foot compared with only twelve dollars a foot in Texas. The American companies were in the favourable position of being able to write off about half of these costs against taxation through depletion and other allowances made for petroleum exploration. But there was the additional high expenditure required to transport the oil, first by pipeline across Alaska to an ice-free port in the south and from there by tanker to California and the West Coast of the United States which was seen as the most likely market for Alaskan oil Most companies worked on an estimate of wellhead production costs of between $1 and $1.20 a barrel not including royalties and taxes, pipeline costs of around seventy-five cents a barrel, and twenty-five cents a barrel for shipping, to the point where it would sell at the market for about $2.90 a barrel. A federal government survey at that time by the Oil Import Task Force set up by President Nixon put the costs much lower, ridiculously so, according to the industry, estimating a wellhead production cost of thirty-six cents a barrel and forty-five cents for transportation by pipeline. If this were so, North Slope oil would be competitive with the Middle East where wellhead production costs are about twenty cents a barrel. Whatever the economic arguments, the companies had to decide on the kind of amounts that could be bid to secure leases on the North Slope and still make a profit on the operation.

A factor in this would be the most recent lease sales in other parts of the United States—in the Santa Barbara Channel and off Louisiana in 1968 and off the coast of Texas earlier in 1969. The pattern had been for a steady increase in both the size and number of bonuses offered, reflecting a growing competition for leases as demand outstripped the domestic supply, and culminating in a record $27,400 an acre for a single lease offshore Louisiana. It seemed inevitable that the North Slope sale would follow this trend. A company would work out a figure based on what other companies had offered at these previous sales and then add something on top of that, depending on how badly it wanted North Slope leases and knowing, of course, that BP and the Atlantic Richfield–Humble group were already sitting on the cream acreage.

As far as the relative value of the different leases was concerned, there was a much greater amount of geological information available than might have been thought from all the scurrying about by the oil scouts and the flavour of international espionage they gave to the whole proceeding. The US Geological Survey material was available to anyone who wanted to see it, as well as the reports of the US Navy's exploration programme. Many of the companies had of course carried out their own geophysical surveys, and even those who had not actually drilled a well could relate those results to the information that had been announced, meagre though it was. And if this were not enough BP, seeking to raise dollars for its own operation, had put up for sale the results of seismic surveys it had carried out, available to anyone willing to pay to see them. One company that did take advantage of the offer was Standard Oil of California.

Before the sale began, the joint holding of leases by various companies had already been well established. Atlantic Richfield for instance held some leases by itself, some in partnership with Humble, and others jointly with BP. The British company also held leases on its own and others with Union Oil. These arrangements enabled companies to spread their risk money over a wider area. What the sale immediately revealed was that during the weeks leading up to it, companies had been actively negotiating with each other to form many new groups. One of the major American companies markedly absent before was Gulf Oil. Now it sought rather belatedly to gain a foothold on the North Slope by coming to a joint arrangement with BP, putting up most of the money in return for BP's expertise. Shell, another previous absentee who had been left on the sidelines by the fast turn of events, went in with Texaco. Mobil, Phillips and Standard of California mostly bid together. And in order to stand a chance in competition with the majors, most of the smaller companies also banded together in groups, the main ones being Amerada-Hess and Getty; Hamilton Brothers, Union and Pan American; and Continental, Sun and Cities Service. One way or another, nearly every American oil company of any size was represented, together with many Canadian firms and the Italian government-controlled organisation, AGIP.

The very first bid opened at the sale brought gasps of astonishment, then wild cheering from the audience that set the pattern for the rest of the day's excitement. BP–Gulf offered $15·5 million for one of six lots on the Colville River delta, a hundred miles north-west of Prudhoe Bay where no major discoveries had ever been made. The bid worked out at an average of just over $6,000 an acre. Excitement mounted when BP–Gulf won the second lot, out of twenty-one separate bids, for an average $8,000 an acre. The third lot went even higher to $12,000 an acre. By the time the BP–Gulf partnership had swept the board by taking the entire six lots in the Colville River delta, amounting to 15,360 acres, they had paid out a total of $97·7 million which was just $100,000 more than the state of Alaska had received in all its previous twenty-two lease sales over ten years. And this was only the start. When the bidding moved to the leases that were available around Prudhoe Bay, prices went so high that even the BP–Gulf coup paled into relative insignificance.

The most spectacular moment came when the bids were opened for tract 57, the closest to the oil discoveries that had already been made. The first bid of twenty-six million dollars was made by Atlantic Richfield. The second by BP–Gulf upped this to $47·2 million. This stayed as the highest for a time as bids of $36·8 million and $36·6 million were announced from the Hamilton group and Continental–Sun–Cities Service. Then the auditorium shook with a roar of shouting and cheering when the Mobil–Phillips–Standard of California group came in with $72·1 million. It seemed that this must be the winning bid. But a moment later, the Amerada-Hess–Getty bid was opened to reveal a figure of $72·3 million, a total of $28,233 an acre. There was a moment of stony silence and sheer disbelief, not only at the size of the bid which was the highest ever offered at any sale, beating the previous Louisiana record, but at its incredible closeness to the previous bid. Then pandemonium broke out. Had there been a leak over what the Mobil–Phillips–Standard group planned to bid? Why had the Amerada group bid alone this time when most of the day they had been bidding in another combination with Louisiana and Land Exploration, Marathon and H. L. Hunt interests?

There had never been any doubt that the tract contained oil. All the companies that had drilled in the area made a bid for it, their eight offers totalling $287 million. One estimate put its recoverable oil reserves at 200 million barrels. Standard of California had taken advantage of BP's offer to sell the results of its seismic work, and so had a good idea of the tract's potential. In addition, Standard of California had been involved with Amerada-Hess in joint bidding for Louisiana leases some months earlier, and so each group knew the way the other was thinking. They had put a similar value on the oil that could be produced, related this to the Louisiana sale, and added a bit for luck. It was a coincicence that they were so close, but no more than that.

The Amerada-Hess–Getty bid was by far the highest winning bid of the day and the group also spent the most, gaining five tracts in the Prudhoe area for $243·5 million. In combination with other interests, the group paid a total of $271·9 million for eighteen tracts. Union and Standard of Indiana came next with $173·6 million for twenty-six tracts. Then the six tracts won by BP–Gulf for $97·7 million and five by Mobil–Phillips–Standard of California for $96·5 million. Other successful bidders are shown in the map on page 169. Altogether, the companies put forward 1,105 individual bids, laying on the table a total offering of $1,676 billion. When the state came to review the results of the sale they accepted high bids on 164 tracts, rejecting fifteen as being too low. The sale finally netted $900,040,000, not quite reaching the magic one-billion-dollar mark, but still the highest in history, averaging $2,180 an acre. It was a far cry from the average of thirteen dollars an acre obtained in the previous three sales only a few years earlier, when in fact 131 of the leases that had just been sold for such breath-taking sums had been put up but received no offers at all. Such was the speed of Alaska's transformation into potentially the biggest oil-producing state of America.

Throughout the day until the sale ended at 5.30 pm, with a short break for lunch, Alaska had become richer at a rate of over two million dollars a minute. Certified cheques for 20 per cent of the amount had to accompany the bids, the rest being payable within two weeks. Since the total interest value alone would be $196,000 a day, not a moment was lost in banking the money. A plane was standing by at the end of the sale to fly the cheques on accepted bids to the Bank of America in San Francisco, while bankers were on hand to cancel other cheques which had been made out by unsuccessful bidders. Not all the bids had been at a record-breaking level, some being won for as little as twenty-five dollars an acre. What had been tipped as the hottest block of all, tract 36 close to BP's discovery well, was picked up by Mobil–Phillips–

Standard of California for only eighteen million dollars, double the only other significant bid made by Humble. BP made no offer for it at all. The sale had its moments of humour as well as excitement. Joke bids of a few cents an acre were made on almost all of the tracts. On one whole tract in fact the highest bid was a dollar, entered by a local Alaskan group who were most incensed at its rejection. The entry into the auditorium of a mysterious figure dressed as an Arab, complete with flowing robes, head-dress and dark glasses, caused a buzz of speculation that a rich Middle East sheikh might be entering the fray—until it was discovered that he was a local humorist in disguise.

As soon as the sale was over, officials and oilmen adjourned to Anchorage's exclusive petroleum club. With the shroud of secrecy no longer necessary they could talk freely together for the first time in months. State officials were still bedazzled by the $900 million bonanza which would take Alaska out of the red for the first time in its history. How to spend it was another matter. Everyone had their pet scheme or project and it was going to be a source of unending argument in the state legislature for years to come with pleas for investment advice going out to bankers and economists all over America. The oil companies too reckoned they had made a good bargain, which became more apparent within a few days when Mobil revealed for the first time that five of its eight exploratory wells drilled earlier that year on the North Slope had struck significant quantities of oil, and BP followed up by announcing that its acreage alone at Prudhoe contained at least 4,800 million barrels of oil, upping the field's total reserves well beyond the original estimates.

But not everyone was so delighted with the sale. Up at the University of Alaska near Fairbanks there is an organisation called the Institute of Social, Economic and Government Research, a dry name for an anything but dry team of dynamic young economists and intellectuals under Dr Victor Fischer, whose job it is to keep a watchful eye on the policies adopted by the state. So wide-ranging are the subjects studied and reported on, from economic relations between Alaska and Japan to the artistic and cultural life of Alaska, that it is popularly known as the 'Institute of Everything'. The opinions of the staff researchers are not always to the liking of government and industrial leaders, but they can sometimes play an important part in the policies adopted by the young state. Since office space at the university is limited, they have their quarters off the campus above the College Inn grocery store which somehow fits in with their down-to-earth, almost impish independence. Gregg Erickson, for instance, one of the Institute's 'young tigers', as he was called in a magazine article, felt that the state had rushed too quickly into leasing too much land to the oil companies—and did not hesitate to say so in word and print. He claimed that the actual worth of the 451,000 acres leased was between two and four billion dollars, and that in addition the state could raise the severance tax on production to 85 per cent and still leave the companies with a better than 10 per cent return on their investment. The companies would no doubt disagree but they have a healthy regard for the astuteness of his economic arguments.

Someone else who also believes that the state is selling itself short to the oil industry is Edgar Boyco, former Attorney General of Alaska in Walter Hickel's administration. He has now returned to his law practice in Anchorage where he continues to prod the establishment in articles and speeches, his opinions sometimes quite outrageous but invariably stated with wit, sincerity and more than grain of truth. He is a spry, smallish man with a wicked glint of humour in his eyes. He has little time for the present state administration: 'It's not that I'm so terribly smart, it's because most of the people in power are so dumb.' He feels that there should be more opportunity for the smaller independent companies and for individuals to share in the oil bonanza—although the way things have developed since, they may feel they are well out of it all—and that the state's royalty from oil production should be much higher, up to 30 per cent.

'I'm not anti-oil company,' he says, and means it. 'They're in business to get the most they can and we need them here. But we must not let them develop our country on their terms. It's like a used-car dealer, if you go on his lot and he shows you this mint condition 1954 Ford and says that'll be $1,500 and you say, yes, I'll take it, and he's prepared to sell it to you for $450 cash, you know, but if you're willing to give him $1,500 he's not going to argue with you or beat you over the head with a bat. But this is what the state of Alaska is doing. We're just like a bunch of hicks. Like a big ship being run by a bunch of weekend amateurs because the professional crew have all gone overboard.'

Boyco—who writes a regular and highly controversial column in *This Alaska* magazine under the heading 'Roar of the Snow Tiger'—is one of those defiant, independent-minded, obstinate, humorous Alaskans who make this young state such a refreshing contrast to the cynicism and world-weariness of much of the lower forty-eight. His answer to what happened to his political career when he left after a year as Attorney General is typical.

'I made the mistake of thinking, well, I've stepped on a lot of toes and if I go back and hang up my shingle as a lawyer it's a bit like a tough cop giving up his gun and his badge and going back to his old neighbourhood to open a grocery store. Evey hood in the district is going to stop by and try to rub him out. I figured I needed some kind of protection so I filed for the state Senate and I made it through the primary. But in the process I suddenly found out that instead of gaining strength as a political candidate I was vastly more vulnerable. Every Tom, Dick and Harry told me what to do and what to think. To give you an example, in the midst of the campaign we had a little experience here in town where a bunch of youngsters who had worked as fire-fighters for the Bureau of Land Management were waiting for their paychecks which the government was slow in processing—so they were camped out on public property near the rail yards and some of them had long hair—only one or two really but they were all bunched together as hippies. They were unceremoniously rousted by the city police and arrested for vagrancy and I came to their aid. It was put in the paper and—oh, man, the storm . . . I lost the barbers' vote and I said to hell with it—I'd nothing against the barbers but if I had to become a prisoner of this group or that group then I'm weaker than I ever was. I decided to be active by assisting the best candidates or those who were less objectionable, but it would be a long time before I'd be a candidate again.'

After the big lease sale, Boyco tried to file an injunction against the issuance of some of the leases but he was predictably not successful. Apart from a small minority, most Alaskans were delighted with the way the sale had gone and at the wealth that was brimming the state's coffers. Most white Alaskans, that is. The only incident that had marred the great occasion was a self-conscious, rather forlorn demonstration by a group of Eskimos and their sympathisers, occasionally interrupting proceedings in the auditorium. Amid all the excitement it seemed an insignificant protest and was soon forgotten. But it was a symptom of seething discontent felt by the natives of Alaska, the original inhabitants in fact, that was to be a vital factor in the oil development that looked so promising in the fall of 1969 but which, by the following year, was beset with problems and uncertainties.

PART THREE

The Environment

14

Death of a Culture

The image many white Alaskans have of an American Eskimo is that of an unwashed drunk, too lazy to work, ungrateful even for the government's welfare handouts on which he lives. Just a few might feel a twinge of guilt at the fate of these once proud hunters who roamed the Arctic wilderness and speared the mighty whale from frail seal-skin canoes. The other image is the tourist one, of happy tambourine-thumping blanket-tossing natives, dutifully performing their ceremonies in a quaintly ramshackle village where, it is whispered, the practice of wife-swapping still goes on although disappointingly they don't seem to rub noses any more.

'Eskimo culture? Hell, the Eskimo never had any culture, except one of survival. He's a vanishing race and the sooner he goes, the better.'

The speaker is a burly Texan, more quietly spoken than most. He isn't an obviously aggressive man. Things weren't easy for him when he first came to Alaska fourteen years ago. There wasn't much money around, work was hard to get. But all that changed when the oil industry moved in. He runs his own construction company now, servicing the drill-sites. When he's really wealthy in a few years' time he plans to get the hell out of Alaska. Meanwhile he's proud to be an Alaskan. He has nothing but contempt for the Eskimos and Indians.

'Take a look round the bars—you'll see what I mean.'

In Anchorage they're on Fourth Avenue, in Fairbanks down by the Chena River. Big tawdry saloons, counters slopped wet with beer, tinny music blaring from juke-boxes. And sure enough, on any night of the week, there are drunken natives, lurching from one saloon to another finally collapsing in the gutter. Like Charlie.

Charlie comes from one of the remote Eskimo villages way up beyond the Arctic Circle. By the time he was seven he had helped to butcher a whale and tasted sour meat at the great whaling feast that followed, not liking it much but enjoying the ceremonial fun and games of the adults. He had seen his grandmother die and watched his parents make love, for they all lived together in one room. He went to school for a while, but it didn't make much sense. He was taught to read about two 'gussuk' (white) children, Dick and Jane, who played with a dog called Spot. Who had ever heard of a boy and girl playing together, or of a dog who didn't do any work and was allowed in the house? And then there was this father who came home each day from a strange place called an office but never seemed to bring any meat with him. School didn't last very long, but at least Charlie learned haltingly to speak and write English, which was more than most of the adults around him could do. He helped his father for a time, fishing for salmon and Arctic char, hunting for caribou and bear on land, whale and walrus at sea. It was easier now than in the days the old people talked about. They had guns instead of bows and arrows and harpoons, outboard motors for their oomiaks, snowmobiles in place of dog teams so that they could range thirty miles to hunt and kill an animal and tow it back to the village all within a day where it might have taken a week before. And the machines didn't have to be fed when they weren't working, which used to be the case with the ever-hungry dogs. No one in the village kept a dog team any more. White men weren't strangers to Charlie. There were the teachers and missionaries, visiting officials from a government agency, even an occasional tourist dropping by in one of the planes that had been a common sight ever since he could remember. Nearly every village was on at least a limited air-service schedule.

The time came when Charlie decided to go south and see the white man's cities for himself. He came to Fairbanks and got a job as a labourer with a road gang and earned good money for a while. Then the season ended and there wasn't any more work. They told him he wasn't educated enough for anything but labouring jobs. He and some new-found friends from an Indian village got together to form something called a trade union which one of them had heard about. Then some men in uniform came and warned them to forget the idea or they'd be in trouble—and they weren't at all like the funny, friendly policemen in the Dick and Jane books.

It wasn't long before Charlie learned about welfare. You just lined up at an office—so that was what an office meant—and they paid you money. It wasn't as much as you could get from working, but there wasn't any work. There was plenty of time though, and not much to do with it except to drink. Everything else was so bewildering. The words he didn't understand. The way people looked at him in the street. There was the girl, of course. She had two children when he met her, now they have another two of their own. They might have got married, but she can get more welfare money by living supposedly on her own with four illegitimate children. With his own welfare money and occasional odd jobs they can just barely survive. Like the one in eleven adult natives who depend entirely on welfare, averaging $127 a month per family. Like the majority of native families who live on $2,000 or less a year, with more than half of those who can and want to work being jobless for most of the year, only one in four having continuous employment. Having lost one culture and unable to assimilate another, drink is the final refuge for the despair of Charlie and others like him. He is not yet twenty-five but looks an old man. The days of hunting and the clean air of the Arctic seem to belong to another age. On average statistics, he won't live beyond 34·5 years. That is the average age of death of native Alaskans, less than in India or Turkey. They have half the life expectancy of other Americans, although sharing with them full citizenship and the right to pay taxes, when they can, and be called up for military service.

Charlie is not typical of all native Alaskans. Some have become skilled airplane pilots, others have taken degrees at universities outside the state and have returned as teachers. A few—very few—have got jobs on the North Slope oil-rigs. The companies are not very keen to employ them, however, for at the hint of a whale hunt or a caribou migration they are likely to disappear for several weeks. The hunting instinct is still very strong. Some natives own businesses and stores and both Eskimos and Indians are represented in the state government. There is the nucleus of a native intelligentsia at the University of Alaska in Fairbanks, where, for instance, a language workshop has been established to formulate written Eskimo for the first time—at present it is only spoken, and then in a variety of dialects—and to record before they are forgotten the ancient Eskimo myths and legends that were previously handed down by word of mouth. (No easy task with words like *ganereyaramtrek* for 'language'.) But they are the minority. Only a few manage successfully to make the transition from one way of life to another.

And not all white Alaskans have the Texan's disdain for the natives. To Claire Fejes, for instance, one of Alaska's leading artists, the

Eskimos are a gentle, gracious people with a nobility of spirit that has largely been lost by industrialised man. With remarkable courage and determination she has on several occasions left her comfortable Fairbanks home, her husband and two children, and gone out into the Arctic wilds to stay for months at a time in those few remote villages where Eskimos still live as they have always done. Her paintings and writings vividly illustrate the simple existence of these nomadic people, harsh, dangerous, sometimes threatened by starvation when the hunting is poor. But such sympathetic understanding is rare. And it is doubtful in any case whether Eskimo communities such as Anaktuvuk and Point Hope can live in relatively undisturbed isolation for much longer.

'A great contrast exists today between the high income, moderate standard of living and the existence of reasonable opportunity for most Alaskans and the appallingly low income and standard of living and virtual absence of opportunity for most Eskimos, Indians and Aleuts of Alaska.' Propaganda by native militants? No—the introduction to an official report by the Federal Field Commission for Development Planning in Alaska, the body which advises on Alaska's economic development. It refers to the fact that most of the 55,000 non-white Alaskans, one-fifth of the total population, live in extreme poverty. Seven out of ten adults have less than an elementary school education. Most live in dilapidated houses under unsanitary conditions. Because of this and unbalanced diets they are more often victims of disease and their life span is much shorter than that of other Alaskans. Infant mortality and suicide rates are twice as high. Mental illness and alcoholism have reached alarming proportions.

The situation is at its worst in the larger villages, where natives tend to drift on leaving traditional ways of living in the smaller settlements. This is where they come into contact to a much greater degree with the trappings of American culture—the movie house, canned goods, the juke-box, even television, for which videotapes are flown in and transmitted from local stations. And since there are few jobs to give them an income to pay for these things, the uneasy conscience of the white man provides them with welfare instead, in the form of unemployment dole and family allowances. It is the worst of all possible worlds, blunting their traditional skills, creating vague longings for unreachable goals, reducing the native to a workless drone living on charity. The result can be seen in a village like Barrow.

Barrow is the largest Eskimo community, at the most northern point of the North Slope. The 'Top of the World', say the signs and postcards provided for tourists, with little regard for geographic accuracy. But it was from Barrow that some of the most famous polar expeditions departed to cross the Arctic ice to the North Pole. Some 2,800 Eskimos live here. There is a hospital, a school and a number of churches of different denominations for this was the first centre from which the early missionaries began to compete for Eskimo souls. One of the results of their zeal was that whereas before the Eskimos had always lain their dead above ground and themselves lived partly below ground in dugouts covered by peat sod roofs—the ice-block igloo was purely a temporary shelter built when caught in a storm while hunting —the missionaries insisted on a reversal of these methods. The Eskimos dutifully buried their dead and built their huts above ground where they could catch the full force of the winds and blizzards which blow constantly across the North Slope. 'I haven't been warm for fifty years,' says an aged Eskimo hunter.

As a result of the 1940s petroleum exploration by the US Navy there is a piped fuel supply to Barrow from a nearby gas-well. Of less benefit are the oil-drums, garbage and discarded equipment they left behind, which are still strewn over the land around Barrow. For debris survives here longer than anywhere else in the world. The Arctic is like a giant refrigerator where the decomposition of matter is extremely slow and metal or plastic is practically immortal. The Eskimos themselves have inevitably contributed to this. In the past, when they were nomads roaming over vast areas, their waste was scattered so far and wide as to be virtually indiscernible. Now, gathered into large permanent settlements where no adequate disposal facilities have been provided, it piles up in huge unsightly heaps. It is not only an aesthetic problem. Coils of wire, jagged pieces of metal, oil seeping from only partially empty fuel-drums can still kill wildlife many years after the departure of the men who left them.

In place of the peat huts there are now ugly prefabricated shacks and squalid one-room buildings made of driftwood and animal skins. There is no proper sewerage system. Most households use pots or pails and dump the waste later on the sea-ice or in nearby lakes, from where the water is also hauled by bucket for storage in the ubiquitous oil-drum until needed. An outside privy is a status symbol; about one in four houses have them, but most of these are unsanitary. In winter, when mercifully covered with a thick blanket of snow, Barrow looks clean, even picturesque. In the short summer, when the garbage-strewn ground is revealed, it looks like a shanty town through which a hurricane has torn. At any time of the year it seems to be populated almost solely by children and old women. These seem to be the only people one sees in the streets. The young children play happily enough, looking well dressed and fed, for the Eskimos have a deep love and feeling for children. The older ones look surly and bored as they rove around in gangs—juvenile crime is high. The old people look merely fatalistic, walking with the bandy-legged roll that comes of soft bones and tuberculosis. Most of the adults are indoors, either sleeping off hangovers or drinking themselves into new ones. Alcohol, fast becoming a major cause of death and suicide, is the great tragedy of the Alaskan native. But it is the result of poverty and loss of pride, not the cause.

In the café at the centre of the village, where parka-clad teenagers gather listlessly round the juke-box and men and women sprawl over the tables in a drunken stupor, there is a poster seeking donations for a fire engine. Fire is one of the great hazards of the Arctic, which seems strange until one remembers that the atmosphere is dry and most of the time the winds blow hard and that water supply is another problem when for most of the year the ground is frozen concrete-hard. The oil companies are well aware of it. At their camps further along the North Slope every building has an extinguisher of some kind and fire regulations are strictly maintained. But Barrow has no fire engine, other than a decrepit tracked vehicle left over from the Second World War. Every winter sees some of the wooden shacks go up in flames, one of the reasons why accidents are the biggest cause of death amongst native Alaskans. There is no way of putting them out, and so the town is saving up for the $25,000 required to buy a fire engine. So far in nickels and dimes they have collected something over $6,000. It makes a bitter contrast with the millions of dollars being spent by the oil companies in providing every possible comfort for their drilling crews, and with the Hilton-like luxury in which American scientists live at the US Naval Arctic Research Laboratory a couple of miles out of Barrow, where, amongst other projects, they are trying to see if dormancy can be induced in polar bears in the way that it exists naturally in other kinds of bear. The contrast becomes even ironic when one considers that the North Slope Eskimos have a legal and moral claim over the land on which such a great wealth of oil has been found.

The conditions in most of village Alaska, the 178 predominantly native places of twenty-five or more persons which are scattered throughout the state with a population of about 38,000, are nothing short of atrocious. Visiting Nome in 1967, the director of the US Office of Economic Opportunity said of its housing: 'Most of the houses are ramshackle, falling-down in places. But even this city has a slum that is worse than the rest of the town where five hundred natives live in the most abject poverty that I've seen anywhere—including Africa, Latin

America, India or anywhere else.' A Commissioner of Public Housing said after a visit to south-western Alaska: 'I've never seen anything like it, even in the worst slums of our major cities. In the lower forty-eight we are trying to get rid of our privies. In Alaska we feel it would be a great advance just to have privies.' It is a terrible indictment of the neglect and indifference shown by the world's most affluent nation towards the native peoples over which it assumed a responsibility when it purchased Alaska a little over a hundred years ago. It is a not unfamiliar story in the early pioneering history of the western United States. What is different is that it happened so recently in Alaska, long after the last of the buffalo had been slaughtered and the few remaining Indians had been herded on to reservations. It is only in the last thirty years that the white man has encroached to any appreciable extent on the lands of native Alaskans where they had been left alone to pursue their traditional means of living by hunting and fishing, their culture largely undisturbed. In fact, until just before the Second World War, they were still in the majority in Alaska, comprising more than half of the total 65,000 population.

But old attitudes die hard, even in a supposedly enlightened age. As late as 1950, there were some native Alaskans still living in virtual slavery. These were the few hundred Aleuts of the Pribilof Islands who were being paid in kind for hunting the valuable fur seal, just as the Russians had done two centuries before—two pounds of coffee a week, a couple of cans of corned beef, a cigar or two, with a cash bonus of no more than $500 a year for only the most successful hunters as an encouragement to better production. Families lived in true 'company' style in houses provided for them, all identical and painted the same colour. The whole area was closed off; no one could visit the islands without a permit from the US government which contributed to the lack of outside knowledge about what was happening there. It took a belated act of Congress to provide the Aleuts with a proper wage for their work, which was worth many millions of dollars a year to the US government, and it was not until 1966 that the area became de-restricted. One man who helped to lead the campaign over the Pribilof Islands scandal was Howard Rock, the sixty-year-old Eskimo editor of a small native weekly newspaper in Fairbanks, the *Tundra Times*.

'There's real poverty out in the villages,' he says. 'But our people are good workers when they're given the chance. Take Point Hope, where I come from. There are about 360 people there, and the only ones on welfare are widows and old people.'

The very word 'welfare' is a sore point with him. It implies laziness on one side and contemptuous patronage on the other. It leads to the kind of attitude revealed in a letter which he recently published in his paper, written to an attorney who is trying to help the native cause.

> Dear Mr ——,
>
> You make us sick working for the lazy, dirty natives. Perhaps you want to line your pockets. Don't see Russia or Canada helping them out. They put them to work.
>
> Do you expect us whites that raised our families here to keep Alaska going and clean up after these natives? I think the decent people will move out and leave you with natives. See what happens then.
>
> Mrs Disgusted

An extremist opinion, whose echoes can be found in London and New York and many other places. Factually wrong even, for both Russia and Canada are making great efforts to help their native populations. But it does reveal the kind of thinking that is creating a widening gulf between white and non-white Alaskans. And the blame for much of this is misplaced welfare. The millions of dollars spent over the years by the federal government on health, housing and education programmes for Alaskan natives have had a marked lack of success; meanwhile, the doling out of unemployment money has increased the sense of despair amongst the natives and aggravated the more ignorant-minded white Alaskans. Very little has been done to find out what the Alaskan native really wants or even to help him help himself. Of the total forty-three million dollars spent in 1968, little more than half a million dollars was for adult vocational training programmes. And even less has been done to set up the kind of industries that might be adapted to particular native skills. As Howard Rock suggests: 'Why don't they set up a tannery in the Pribilof Islands, instead of sending all the skins down to St Louis for treatment?'

As with most other things in Alaska, it was the arrival of the oil industry that brought these problems most sharply into focus. On the one hand it can provide employment and opportunity for Alaskan natives, not only on the actual oil-rigs but in the local services required, such as transportation, supplies and provisions, construction and many others. On the other hand, the drilling operations of the industry on the North Slope and elsewhere and even more the projected pipeline to be built from north to south across Alaska will inevitably have an adverse effect on the traditional hunting grounds of the Eskimos and Indians, reducing the amount of game available to them. And this precisely represents the dilemma of the Alaskan native as he stands poised between two cultures; how much to retain of the one and how much to accept of the other. It is in fact the dilemma that faces all Alaskans. Oil can bring the advantages and opportunities of economic growth; the price that has to be paid for that is a degree of encroachment on the wilderness. But while the conservationists and businessmen argue in terms of aesthetics and financial loss or gain, to the Alaskan native it is a matter of sheer survival.

Oppressed as he is by poverty and despair, the 'Eskimo Power' stickers that can be seen on automobiles in Anchorage or Juneau seem to have a hollow mockery about them. But in fact that power is very real, of which the natives of Alaska have become increasingly aware. It stems ironically enough from the very neglect that they and the territory as a whole suffered in the past from succeeding governments in Washington. No Congress ever bothered to pass a treaty or come to an agreement over the question of the title to Alaskan native lands. There didn't seem any need to—no one was very interested or concerned with Alaska. Consequently the Eskimos and Indians and Aleuts were not cheated out of their rights or cunningly persuaded to sign disadvantageous treaties as happened in other western states—the Indians of California sold virtually the whole of what was to become that state for twenty-nine million dollars. That might have seemed to be a fair market value in 1853 and the Alaskan natives would probably have accepted a similar kind of deal if it was offered to them. But it was not. They now had a valid claim over much of the land in Alaska, worth far more because of the oil discoveries than at any time in the past. Until the native land claim had been settled, further oil industry development in Alaska was at a standstill. Delayed justice had for once played into the hands of the Alaskan natives. As one Indian leader put it: 'In our case, justice is not now going to come very cheaply. We have been on the bottom step of the ladder too long.'

15

Deadline for Justice

As long ago as 1869, only two years after the United States purchased Alaska from Russia, the Tlingit and Haida Indians of south-eastern Alaska made their first protest to Washington about what they considered to be the illegal sale of their land. John Borbridge, first vice-president of the Alaska Federation of Natives, which is now fighting the native land claim and himself a Tlingit chief, relates this with some pride and satisfaction.

'The Russians owned and controlled very little of Alaska,' he points out. 'They were settled at Sitka through a process of negotiation and force. But whenever they wanted food or game from the inland regions, they had to trade with the Indians. Their domain was very limited.'

In fact the Russians rarely set foot in the interior, and as for the northern region beyond the Brooks Range, there is no evidence that a Russian ever saw it. Their only claim, and that by right of conquest, was to the Aleutian Islands and part of the coast of south-eastern Alaska. When the United States acquired Alaska, even accepting that it was legal by the standards of those days, it purchased not the land itself but only the right to tax and govern. It recognised that the land belonged to the original inhabitants—the Eskimos, Aleuts and various Indian tribes—and acknowledged that fact in the Organic Act of 1884 which established a territorial government in Alaska, promising that the natives 'shall not be disturbed in the possession of any lands actually in their use or occupancy or now claimed by them'. And this is the basis of today's native land claims. The problem is that no Congress ever got around to passing legislation to convey title of the land to the natives. It didn't seem to be a matter of any urgency and the natives themselves had no cause to think their rights were being threatened in any way. Not until the Statehood Act of 1958 that is, by which Alaska became a state in January of the following year.

Under the terms of the Act, the state was given the right over a period of twenty-five years to select for itself 103·5 million acres of land from the public domain, those federal lands which had not been previously set aside for specific purposes and reservations. Basically it left the state with a choice out of some 270 million acres of unreserved land. In order to protect native land rights, Congress provided that the 'state and its people do agree and declare that they forever disclaim all right and title . . . to any lands or other property (including fishing rights), the right or title to which may be held by any of the Indians, Eskimos or Aleuts'. Nevertheless, the state did move to take over lands which were used or occupied by the natives. The Indians of Minto village, whose lakes in that vicinity provide one of the best duck-breeding grounds in the world, found their land was slated to become a recreation area for tourists and sports hunters. The Indians of Tanacross Village discovered that their lands on beautiful Lake George were being offered for sale at the New York World's Fair as 'wilderness estates'. But the biggest lure for the state was those lands which offered the chance of oil discoveries, and these included the two million acres along the shore of the North Slope. It was claimed that the land was free of aboriginal use and occupancy, although for centuries the whole of the North Slope had been a traditional hunting and fishing ground for the Eskimos. The land was granted to the state by the federal Bureau of Land Management in 1964; later that same year the state held its first oil lease sale there. Few Eskimos knew of the transfer at the time, and even today there are villagers in the more remote areas who are not even aware that there is any question of right to land they have always regarded as their own.

By 1966 the Bureau of Land Management had granted Alaska title to 5·8 million acres of land, tentative approval of a further 7·9 million acres, and the state had selected and applied for a further 12·3 million acres. Many of these lands were the subject of native claims and the natives had become thoroughly alarmed. Appeals were made to Washington, with the result that later in that year the then Secretary of the Interior, Stewart L. Udall, halted the transfer of additional land and suspended the issuance of new federal oil and gas leases until Congress had settled the matter of native land rights. The Governor of Alaska, at that time Walter J. Hickel, who himself later became Interior

Secretary, condemned the act as illegal and the state filed suit against Udall to force him to complete the transfer of certain lands. The case still rests in the federal courts; meanwhile, as one of his last acts in office in January 1969, Udall formalised the 'land freeze' by issuing a public order which stated: 'This action will give opportunity for Congress to consider how the legislative commitment that the natives shall not be disturbed in their traditional use and occupancy of the lands in Alaska should be implemented . . . To allow these lands to pass into other ownership in the face of the natives' claim would, in my opinion, preclude a fair and equitable settlement of the matter by Congress. It would also deny the natives of Alaska an opportunity to acquire title to lands which they admittedly have used and occupied for centuries.'

Some local native land claims go back over thirty years, but the only native organisations that existed before statehood were the Alaskan Native Brotherhood and Sisterhood, made up of Indians from south-eastern Alaska. It was from 1961 onwards, beginning with the Inupiat Paitot of northern Eskimos, that regional groups came to be formed to file protests and land claims and to seek compensation for lands that had already been taken from them. There were twenty-one groups in existence by 1966, such as the Arctic Slope Native Association and the Copper River Indian Association, claiming a total of 290 million acres of land. They had little connection with each other, and nothing had been done to co-ordinate any joint action. It was mainly at the instigation of Emil Notti, Willie Hensley and Eben Hopson, leaders of the northern Eskimos, and John Borbridge and other Indian leaders that all the groups got together in 1966 to form the Alaska Federation of Natives to fight for their land claims as a common cause. All the regional claims and protests were amalgamated and the Federation began the task of negotiating with the state and federal governments over a just settlement of the one overall claim. With the land freeze preventing the state from selecting further land and, at the same time, holding up any new oil operations other than those already started, it was in everyone's interest to come to an agreement as quickly as possible.

'The first offer they made us was for a total payment of $7·2 million —just what the United States originally paid for Alaska in 1867,' Borbridge recalls, with a wry smile at the ignorance and stupidity that the government obviously expected of the natives. That offer was rejected out of hand, as well as another in 1967 of $180 million. It was going to be a longer battle to come to terms than the politicians, eager to get on with the business of developing Alaska, had exptected. There were many willing to volunteer their time and effort to the cause, but it was obviously going to cost money for legal expenses and to obtain expert advice. Financial support came to the Federation at just the right moment and from an entirely unexpected source.

In the village of Tyonek, on the west shore of the Cook Inlet forty miles from Anchorage, live the richest Indians in Alaska. The 280 men, women and children who are the sole survivors of the Moquawkie tribe, some sixty families in all, share the income from a $14·7 million windfall that came to them in 1964 and 1967 from the sale of oil and gas leases on their land. They live in modern ranch-style houses of four or five bedrooms, with every convenience that one would find in an exclusive New York suburb. Their children are educated in one of the most up-to-date schools in the state with a pupil-teacher ratio of fifteen to one; the older ones have a choice of going to any boarding school or university in the United States. Among their investments they own some of the largest office buildings in Anchorage—the utility company that provides much of the city with its water and sewer services, a major construction company and an air charter operation. They handle all the money themselves through a management corporation and a village council, which carefully decides how it should be spent. Requests for second colour TV sets and fancy cars have been turned down; pick-up trucks, furniture and clothing are generally approved. There is no road into the village. The only practical means of getting there is by plane—and the airstrip is closed to all but invited guests. This has kept out the stream of salesmen and insurance and investment brokers who, in the early days, came prospecting for the Tyoneks' wealth. An advertisement placed in the Anchorage papers addressed to salesmen in general said: 'Don't call us—we'll call you. The scalp you save may be your own.' The Tyoneks were determined not to be conned out of their money as Indians in other states had been. They proved that, given the chance, the Alaskan native could manage his own affairs with wisdom and responsibility. The story of the Tyoneks is unique; their accomplishments remarkably successful.

They were not always so fortunate. The Moquawkies were a tiny branch of the Athabascan Indian tribes of the interior, who settled in the marshy, spruce-covered land on the north-west shore of the Cook Inlet and lived by hunting and sea-fishing. Close as they were to those areas of the south coast first visited by white men, they were among the first to suffer from the effects of his invasion. Game became scarce, later the fishing declined too as the salmon-canning industries moved in. When President Woodrow Wilson set aside a 24,000-acre swamp as a reservation for the Moquawkies in 1927, it seemed to be just another bad deal for the red man, giving him title over lands he already possessed and then only because no one else wanted them. It was in fact one of the very few Indian reservations created in Alaska. Other parts of the territory were so remote from white settlements that it wasn't considered necessary.

Things went from bad to worse for the Tyonek villagers. By the 1950s their numbers had dwindled to a couple of hundred. With little hunting and fishing left they were almost entirely dependent on welfare and social security. They were living in the kind of conditions that other Alaskan natives do to this day, in dilapidated tarpaper shacks, often as many as twelve to a room taking it in turn to sleep. In their despair it was hardly surprising that much of their welfare money went on liquor. The Tyoneks had the reputation in Anchorage of being the town drunks. Living only forty miles away, they were the Indians that Anchorites saw most of and were generally despised by the local people. A networked television programme in 1955 called 'People Are Funny' chose the village of Tyonek to show the reaction of the Indians to the delivery of a TV set. 'The amazed villagers babbled excitedly and pressed closer to the TV receiver to get a better look at this electronic marvel,' went the publicity blurb. It was apparently thought a real hoot to show the wonders of modern science to primitive Indians who were living on the brink of starvation.

Not everyone in Anchorage had the same attitude towards the Tyonek Indians. One person who had a deep sympathy for their plight was Stanley McCutcheon, an attorney born in Anchorage in 1917 when it was little more than a camp-site for the railway construction crews. He undertook to represent them legally when he became a lawyer, more out of friendship than expecting any payment. More often than not it involved bailing them out of jail when they came to town and got drunk. It was he who took up their case when, as a result of the oil discoveries on the Kenai Peninsula in 1957 and in the Cook Inlet a few years later, it was thought likely that oil and gas existed on the northwest shore of the Inlet as well, on the land of the Moquawkie reservation. The oil companies were anxious to obtain leases there as a prelude to drilling.

It was a long and bitter battle for the Tyoneks to retain their rights even on land which had been reserved for them. Now that it was possible they were sitting on a valuable oilfield, there were plenty of people willing to find a reason for it not being theirs. The Interior Department claimed that it was only the land that belonged to the Indians and not the mineral rights. A federal land sale was scheduled in April 1963 and only called off at the last minute as a result of lawsuits brought by McCutcheon on the Tyoneks' behalf. When their right to sell the leases was recognised, the Bureau of Indian Affairs tried to persuade them to accept an intial offer of one million dollars for them. It was feared they would receive more per acre than the state had in its own lease sales. In the end the Tyoneks did just that, ironically because of the delays caused by government interference which sent the bidding up higher. They accepted in March 1964 bonus bids totalling twelve million dollars for the 25,573 acres that had been offered and agreements giving them an annual rental of $1.25 an acre plus a 16⅔ per cent royalty on any petroleum produced, as against a royalty on state land of 12½ per cent. A further $2·7 million was brought in at a second sale in 1967 of land where bids had been rejected as too low in 1964. The final irony was that in spite of drilling by a number of companies, no oil was ever found on the Tyonek land. One well found a small accumulation of natural gas which, under their agreement, now belongs to the Tyoneks since there is no commercial outlet for it. This gas now provices fuel and power for the village. Through the strange ways of oil, the Tyonek Indians have in some measure been able to balance the scales of justice against the white man who deprived them of so much in the past.

But even when they had got their money, things were still not easy. There were Alaskans who said they had no real right to it and would fritter it away. The Bureau of Indian Affairs wanted to administer it for them—as if the history of its own conduct was anything to boast about. But eventually the wise leadership and the counsel of people like McCutcheon prevailed and the Tyoneks showed just how well they could manage their own affairs. In the first days after the original sale there were still many people in Anchorage unaware of what had happened. Their treatment of the Indians in the past had not endeared them to the now wealthy Tyoneks, who went to Seattle to order the houses they were to have built.

'But one thing they thought they would get locally was silver-ware,' McCutcheon recalls. 'In place of the tin knives and forks they had been using. One of them was sent to Anchorage to see what was available. He was chased out of the first two stores he visited—they thought he was just another drunken Indian. In the third store he found an eighty-five dollar set. The owner was amused, didn't expect him to buy it. You can imagine his shock when the Indian ordered sixty sets—and wrote out a cheque for them there and then. The word went around that the Tyoneks had money, and people's attitudes changed overnight. Salesmen flocked down to Tyonek in their hundreds, until the chiefs put a ban on outside visitors.'

Bitterness between the Tyoneks and the local Anchorites hardened. People were jealous of their luck and new-found wealth. It took a series of sudden tragedies to alleviate this and lead towards the gradual establishment of better relations to the point where most Alaskans are now proud of the Tyoneks and their achievements.

On the night of April 14, 1966, the last that the Theadore Chickaluson family were to spend in their ramshackle woodframe cabin before moving the following day into one of the new houses in the village, the cabin was destroyed by a fire which killed four young grandchildren of the old chief, Simeon Chickaluson. In an argument a few weeks later Albert Kaloa, Sr, father of the village president, shot and killed a fellow villager. He pleaded guilty to manslaughter and was given a five years' suspended sentence. And then on September 12 his son, Albert Kaloa, Jr, who more than anyone had fought the Tyoneks' cause, planned for their future and given then back their pride, was burned to death in a hotel fire in Anchorage which also killed thirteen other people. It was the final cruel blow that numbed the Tyoneks and shocked all of Alaska. But Albert Kaloa had prepared the groundwork well for his people's future prosperity. They created a village community that was a lesson to the sceptics who said they could never manage their own affairs and a source of pride and encouragement to all Alaskan natives. And in more practical terms, they provided much of the financial support that enabled the Alaska Federation of Natives to fight the overall land claim battle.

Support of another kind came from another offshoot of the Athabascan Indians, the Tlingit and Haida tribe. Over the years, when the individual native groups were making their separate claims, it was only the Tlingit and Haida Indians who had come to an actual settlement. Because of their well-formulated social structure and their longer contact, first with the Russians and then the Americans, they were in a more advantageous position to press their claim to land title over most of the twenty million acres of south-eastern Alaska. They had early on protested at the illegality of the sale of Alaska. Then in 1935, in the first positive action by any of the Alaskan native groups, they sued the United States for compensation over lands taken from them earlier to create, for instance, the Tongass National Forest and the Glacier Bay National Monument. Their right to make this claim was upheld in 1959 by the US Court of Claims and resulted in the settlement in 1968 of $7·5 million in compensation. The Tlingit and Haida Indians considered this to be grossly inadequate but accepted, partly to use the money in helping to fight the much wider land claim issue which had by then arisen but mainly because it set a precedent in allowing the validity of such claims. As one Anchorage attorney wrote: 'It is difficult or impossible to read the decision of the Court of Claims in TLINGIT and

HAIDA without concluding that counsel for the government were either asleep at the switch or resting on their oars.'

With financial resources to back them and a growing feeling both inside Alaska and outside that here there was a chance to see that at least some of America's aboriginal people obtained justice, the Alaska Federation of Natives became even more determined. Land was their main concern, as graphically set out in one of their publications, *Native Alaska: Deadline for Justice.*

'To the Alaska Natives, the land is their life; to the State of Alaska it is a commodity to be bought and sold. Alaska Native families depend on the land and its waters for the food they eat, hunting and fishing as they have done for thousands of years. . . . Often a thousand acres are required to support one person. In some regions a village of two hundred people may require as much as 600,000 acres.

'Their subsistence economy is as varied as the land. The Eskimos of the Arctic hunt whale, walrus and seal in the coastal waters and caribou on the frozen tundra. They gather murre eggs from the sea cliffs in July and take ducks and geese through the summer. The Athabascan Indians of Interior Alaska, on the other hand, hunt moose, beaver, muskrat, rabbit and black bear in forests of white spruce and birch. The lakes offer ducks and geese. At their summer fishcamps they take salmon by fishwheel and salt or smoke it for the long winter. When the waters freeze over, pike are caught through the ice. Wild blueberries are gathered in July and cranberries in September.'

After turning down further niggardly offers from the state and federal governments, the Federation late in 1969 came back with its own claim. This basically called for the title to forty million acres of land including the mineral rights (reduced from the eighty million acres originally sought), cash compensation of $500 million, and a 2 per cent residual oil royalty on gross revenues from federal lands to which native title had been extinguished. Presented as a proposed bill before Congress, it was largely drawn up by Arthur Goldberg, former Supreme Court Justice and US ambassador to the United Nations, who in July of 1969 had agreed to represent the native cause. The 'Goldberg bill', as it was called, came as a bombshell to Alaskans, provoking an angry reaction. An editorial in the *Anchorage Daily Times* called some of the provisions 'shocking', which would 'cripple the development of Alaska for all its citizens'.

Since its formation the Alaska Federation of Natives had taken great care to base its case within the framework of legal precedent in the United States, citing instances where the rights of native occupancy to land had been upheld in other states, and sternly rejected any proposals for militant action which, in any event, would be alien to the temperament of Alaskan natives.

'We were trying to get justice in a legal way,' says John Borbridge. 'We were seeking our aims within the institution of government. As a matter of fact, at this particular time, I think the people of the United States very badly need to see that justice can be done in a case such as ours, whatever the cost. The natives of Alaska have had to adjust to a new situation—so must the non-native Alaskans.'

And so the legal arguments started—and continued until 1972. There is little that is more complicated in American law than Indian land rights, striking to the very heart of the white man's conquest of the nation, and the kind of problems involved had not arisen in quite the same way since 1912 when New Mexico and Arizona had been admitted to the Union. It was argued against the natives, for instance, that when Russia ceded Alaska to the United States the original price was seven million dollars, but an additional $200,000 was paid in return for Russia adding a clause guaranteeing actual title to the land, in which case it would extinguish all native rights to land. Attorneys for the Alaskan natives contested this, holding that the extra $200,000 guaranteed title only against the Russian America Company and was not applicable to natives. Further arguments revolved around the phrase 'use of land' and what was really intended by it; does hunting and trapping apply? The trouble was that no Congress had ever defined just what is meant by 'use and occupancy' of land.

Finally, at the end of 1971, Congress passed a bill which gave the Alaskan natives virtually all they had demanded in the first place—forty million acres of land, a sum of $462·5 million to be paid by the US Treasury over an eleven-year period, and a further $500 million from oil and other mineral royalties. The bill was signed into law by President Nixon early in 1972. It also included authorisation for the Interior Department to select up to eighty million acres as national parks and wildlife refuges. These selections are being contested by the state of Alaska so that while the original 'land freeze' has been lifted, further delays have arisen. Time is running out for the state of Alaska which itself has only until 1984 to select the land it wants. Already, several millions of dollars have been lost in lease sales which have had to be cancelled. Officials have warned that as things are, the state could be bankrupt by 1976; meanwhile much of the oil exploration programme which showed such promise in 1969 has ground to a halt.

The biggest effect of the land freeze on the oil companies was that it precluded the leasing of a right of way along the route of the proposed 780-mile pipeline from the North Slope to Valdez on the south coast, most of which is federal land. It was in fact just one of the factors which have prevented the pipeline from being built. Another also has to do with the natives' concern over land, but this time in connection with a specific group. One of the regional groups of the Athabascan Indian tribes of the interior is the Tanana Chiefs' Conference, representing about thirty villages over a wide area around Fairbanks. And it is through the hunting grounds of some of these villages that the proposed pipeline would be laid.

Leader of the Tanana Chiefs' Conference is Altred Ketzler. In 1962 he was twenty-nine years old, owner of a hardware store in Nenana, the Indian village south-west of Fairbanks in which he had been born. He also happened to be a traditional chief of one of the Athabascan tribes. The various tribal chiefs had not met together since 1913, but with the state threatening to take over land they felt to be rightfully theirs Ketzler thought it was time they got together to consider the problem. He sent invitations out to all the village chiefs in the region and on a bleak day in March, still winter-cold but with the ice in the river beginning to break up, they gathered in Nenana. After traditional celebrations and dog-sled races they got down to serious business. A council was held and it was decided to revive the Tanana Chiefs Conference, the second of the native regional groups to be formed. For the first time in its long history a democratic vote was held to decide on who should be president of the council, instead of following the previous custom of appointing the most senior chief. Ketzler was elected. And so began his involvement in politics which led to his becoming a director of the Alaska Federation of Natives when it was formed in 1966. It meant a considerable sacrifice in giving up his business in Nenana, coming with his wife to live in Anchorage, and travelling constantly all over the United States to seek support for the Alaskan native cause.

'The problem of my people is that they are land orientated,' he says. 'An Indian family needs thousands of acres of land in which to hunt and fish and sustain itself. Land is more important to us than money.'

Ketzler works from the Federation's headquarters in Anchorage, housed appropriately enough in the new two-million-dollar Tyonek office building of which IBM rents a large part. It has the atmosphere of a political or charity campaign headquarters, enthusiastic, slightly amateur, busy with secretaries and volunteer helpers, occasional sympathisers dropping by to offer their good wishes. It is difficult to imagine Ketzler in his role of tribal chief. He wears an ordinary business suit with a CND badge on his lapel. His accent is indistinguishable from any white Alaskan, his slight swarthiness could place him as an Italian. He mentions that his nephew was killed in Vietnam in 1970—not in the front-line fighting but in the hospital where he worked as a medical orderly. He and one of the other medics were shot dead by a drunken GI who went berserk. In letters home before he was killed he wrote that more than half the patients in his hospital came in with wounds inflicted either by themselves or by other American soldiers.

When the route of the proposed trans-Alaska pipeline was first revealed in 1969, it was seen by Ketzler and the other Tanana chiefs to cut across the traditional hunting grounds of five villages north and west of Fairbanks—Stevens, Rampart, Bettles, Minto and Huslia-Hughes—which were included in land claims already filed by the Indians. Rather than seek cash compensation from the oil companies for right of way over what they considered to be their land, the chiefs saw this as a means of providing employment for their people and at the same time start a native industry that could continue after the line had been built. They formed a development corporation to contract for some of the many services that would be required for such a mammoth construction project, similar in fact to companies started or expanded by white Alaskans, particularly in Fairbanks, which were already making a lucrative profit by undertaking contract work for the oil operations on the North Slope. In return for this, the Indians would grant a right of way over their land. But when they came to enter bids with TAPS, the pipeline company, these were ignored. To be fair to the oil companies there was little else they could do. Neither the federal nor the state government recognised that the land belonged to the Indians, and in any case the land freeze prevented the necessary permission being granted to build the pipeline until the whole question of the native land claim was settled. TAPS was caught squarely in the middle of a dispute that was none of its own making. However, that did not stop the native development corporation from suing TAPS for twenty million dollars for breach of contract and to declare as invalid the waives that had been granted to the company. The case still rests in the Alaskan courts.

Of greater significance was a second lawsuit brought in Washington by the five villages concerned, seeking an injunction against the Department of the Interior giving permission for the pipeline to be built across their land without their consent. It was contended by government lawyers that since the line would not actually penetrate the villages their consent was not required. Nevertheless at the end of March 1970, although denying the request of four of the villages, Federal District Judge George L. Hart, Jr, issued a restraining order on behalf of Stevens, a village on the banks of the Yukon River occupied by sixty-six Indians who claim ownership to land that would be traversed by some twenty miles of the pipeline. That injunction, which is still in force, caused great indignation in Fairbanks where businessmen and construction workers alike feared their livelihoods would be threatened if the pipeline did not go through. Already much of the necessary construction equipment had been amassed and was ready to make a start on the road that would first have to be built along the route of the line in order to haul the pipe in place. The Washington court's decision smacked of the kind of outside interference that Alaskans had resented for so long, and at that time the Governor of Alaska, Keith Miller, had unearthed an obscure 1866 law which he considered gave the state the right to authorise highway construction on federal land. This seemed to be a way out of the whole embroiled situation, a loophole in the land freeze imposed by the federal government—and now even that was stymied by a few poor, ragged Indians. A deputation of Fairbanks' businessmen and officials went to Stevens to 'persuade' the villagers to change their minds.

'I was there at the time,' Ketzler says. 'The district attorney and others were closeted with the village council for three hours. I wouldn't say there were outright threats, but there was overt coercion—like they might not get the bigger airfield and other things they wanted unless they agreed to withdraw their lawsuit.'

It was said that the villagers had been misled and wanted the case thrown out of court. The Indian leaders denied this. The somewhat sinister implications involved led to an investigation in Juneau by a special sub-committee of the state government. In the end the case stood, the villagers did not want the injunction withdrawn. But by then it was of less importance. Further complications had arisen to cloud the issue and to delay even longer construction of the pipeline. Conservationist groups had obtained an injunction of their own in Washington against the project. The trans-Alaska pipeline was no longer just a state matter, but had become enmeshed in national politics and the whole question of pollution and environmental control.

16

The Conservation Issue

The North Slope of Alaska, remnant of a retreated ice age, is not the tough giant it first appears to be. Nor is it quite the barren wasteland devoid of life so often depicted in film and photograph. The long Arctic winter is certainly harsh and awesome to men not accustomed to it. They have good reason to fear the fogs and blizzards that can reduce visibility to zero and the great ice packs that crush against the shore and the relentless cold that is not just cold but a tangible presence. But these forces hide one of the most delicately balanced and fragile environments in the world, as evidenced in summer when for a few weeks the caribou in their thousands come to graze on a sudden carpet of moss and buttercups and wildflowers and the air is alive with rare waterfowl and bumblebees are the memory of an English garden.

This was little understood by the oil companies when they moved into the region in the first major attempt by private industry to search for oil in the Arctic. For they came in winter to begin with, prepared to do battle with a frozen, hostile wilderness. It was just another area to explore in the constant need to supply an oil-hungry world, different perhaps and with more formidable problems than anywhere else. But they were the kind of problems that the industry with its technical and financial resourcefulness had always managed to overcome before, whether in desert or jungle or pioneering in offshore waters to find oil and gas beneath the seabed. In a mighty effort dwarfing any other single oil operation, they succeeded on the North Slope. They tamed the last great wilderness on earth. As in any such large-scale intervention by man, there was a price to be paid in terms of its effect on the natural environment. The price was greater than the companies themselves had first realised, as they gradually came to understand the terrible vulnerability of the region. But, they reasoned, since it was inevitable that any environment in which they chose to search for oil would be affected, where could be better than this utterly remote and unpopulated place? No one had ever shown the slightest interest in it before. Only a handful of naturalists had ever been there, most people had never even heard of it.

But far from being congratulated for their technological interprise, the oil companies found to their surprise and indignation that they were being criticised from all sides on grounds of conservation for endangering the ecology of the area. And if they thought it was a passing phase or something that could be overcome by their powerful lobby in Washington, they soon found that the weight of opinion against them was so strong that it resulted in tangible legislation to curb their operations. Court injunctions brought by conservation groups prevented any attempt to make a start on the trans-Alaska pipeline. By mid-1972, when according to the original schedule there would have been around a hundred development wells drilling on the North Slope with the pipeline nearly completed to begin pumping the following year, operations had come nearly to a standstill. Oil-rigs were either stacked or moved away to other areas. Half a million tons of steel pipe, enough for more than the 800 miles required and bought at a cost of $130 million, lay rusting in forty-foot sections at sites along the proposed route. The giant graders and earth-digging machines assembled by Alaskan companies in a gamble that the necessary construction permits would be given, remained still and unused. For the time being, the oil boom was over. For the first time in its history the oil industry was confronted with a problem that it seemed unable to overcome, not brought about by the forces of nature but by the vagaries of human emotion and a feeling that the price demanded for industrial progress might be getting just too high.

This general awareness of the need for conservation, bringing with it a new language of 'ecosystems' and 'environmentalists', was not confined to Alaska or even to the United States. It was felt by many people in all the advanced industrialised nations who, for the first time, began listening to what scientists had been warning for many years. In one sense it was a product of affluence, for it is difficult to interest a starving man in the need to limit the spraying of crops by pesticides, or to tell those living in tenement poverty that houses should not be built in a beautiful stretch of countryside. The argument easily becomes one of the haves and have-nots. But there were particular reasons why the controversy should focus on Alaska; why Alaska, in fact, became the major battlefield in the whole issue of conservation, representing all the shades of opinion and environmental problems that confront the rest of the world.

The neglect which Alaska suffered in the past was bitterly resented by Alaskans who wanted to develop their territory so that it could stand on its own feet, but it also had its advantages. While the march of progress swept across the rest of America in the wake of a pioneering eagerness to tame the wilderness, Alaska remained reasonably intact, a haven for the lone trappers and gold-miners and a few rugged individualists. The northern wilderness was forbidding enough to deter even the cat-skinner (bulldozer driver) who had become the modern David against the Goliath of nature as he tore up forest and scrub to make way for civilisation.

In Alaska the frontier still remained, not a mere echo of the past but a very real presence, much as California was at the end of the last century or Texas fifty years ago. The gradual awareness of this by people in the lower forty-eight, some of whom moved northwards for that very reason, began at a time when the oil industry decided to explore for oil in northern Alaska. There had always been a few naturalists and others concerned with conservation who had spoken out against the increasing despoliation of natural surroundings, but in the main they had been ignored or dismissed as cranks. Now, with the results of pollution visible all around them, people began to listen. Two events brought home the message in a way that no amount of propaganda could have done. First there was the *Torrey Canyon* disaster in March 1967, when a large tanker ran aground off the Cornish coast of England causing severe pollution to beaches normally used by holidaymakers and killing many thousands of fish and sea-birds. And then, in 1969, the beaches along the Santa Barbara coast of California were similarly polluted as a result of widespread oil leaks from offshore drilling-rigs. In fact, such dramatic examples of pollution are not as harmful in the long-term as the less publicised but continuing pollution caused by tankers washing out their empty tanks at sea, inevitably increasing as more and more oil is moved around the world to satisfy a seemingly insatiable demand. But they led to a public outcry, particularly in the countries most affected. Even this might have died down as the beaches were cleaned and returned to normal. But then President Nixon's administration decided to make conservation a major political issue, motivated no doubt by the groundswell of public opinion but also as a means of diverting the mind of America's rebellious youth away from the Vietnam war.

As politicians jumped on to what appeared to be a most promising bandwagon, they vied with each other in demanding ever more stringent restrictions against pollution, particularly aimed at the oil industry. Lease sales in new offshore areas were cancelled, production was halted or severely restricted in existing producing areas, there were even attempts to take back some leases already sold by the government off the Santa Barbara coast. The acquisition of sites for new refineries, power plants and tanker terminals was held up on environmental grounds. Restrictions were placed on the amount of sulphur fumes that could be emitted from power stations, penalising coal and fuel oil imported from Venezuela which is cheaper but also has a high sulphur content. There were moves towards legislating for a lead-free gasoline. The political drive towards protecting the environment and cleaning up pollution culminated in the National Environmental Policy Act of 1969 and the National Air Quality Standards Act of 1970. Under the first, no industrial projects can be approved until public hearings have been held to allow conservationists the chance of opposing them and, if their case is strong enough, to defeat them. Under the second, the toughest anti-pollution laws ever passed by a government, new control standards for clean air and water cover every segment of industry, including a requirement for automobile manufacturers to reduce harmful exhaust emissions from cars by 90 per cent by the beginning of 1975.

The need to clean up the environment certainly existed after the many years in which American industry, unlike its more restricted counterparts in Britain and Europe, had been able to do very much as it wanted by way of pollution with succeeding governments showing almost complete indifference to the whole question of conservation. Now, within five years, American companies will have to spend some twenty billion dollars to clean up the atmosphere, streams and rivers in order to conform to the legal standards established or face heavy fines or even imprisonment. It is estimated that seven industries will have to foot clean-up bills of more than one billion dollars each, with electric utilities at the top of the list with over three billion dollars, then iron and steel with over $2·5 billion, and thirdly the petroleum industry with some $2·2 billion.

Coming all at once in a short space of time, the environmental and pollution restrictions have caused traumatic upheavals within the energy supply industry. Heavier reliance on cleaner fuels from oil and natural gas caused an upsurge in demand at a time when both were in short supply. With rising tanker freight rates caused by the continued closure of the Suez Canal and drastic price increases demanded and obtained by the Middle East producing countries, imported oil is no longer cheaper than domestic oil, even apart from considerations of national security by too great a reliance on imports. The position of natural gas is even more critical. Before the Second World War it was regarded virtually as a waste product, burned off in oilfield flare-stacks where produced as a by-product of crude oil. Because of low price levels enforced since 1954 by the Federal Power Commission, which controls the industry as rigidly as any nationalised European organisation, demand has increased to the point where natural gas is nearly equal to oil as an energy fuel and provides a third of the total energy consumed in the United States. At the same time low prices have, in many instances, made the search for new sources of natural gas uneconomic so that reserves have been steadily falling over the past three years, equivalent now to only eleven years' production at current rates of demand. But as the cleanest fuel available to meet anti-pollution requirements, for electricity generation for instance, the demand for gas is increasing even more rapidly and has now reached the point where it cannot be satisfied by domestic production alone. Supplies are being imported by pipeline from Canada and plans are in hand to import considerable quantities from North Africa, in liquefied form under pressure in special tankers, and also to make synthetic gas from oil and coal. By the end of the 1970s more than half of the gas consumed will have to be imported or made from other fuels.

These additional sources of supply are not yet available, however. It will take time to build the necessary tankers and to develop the technology and to build the plants to convert other fuels into gas. The United States therefore faces an energy gap, and the shortages which already began to cause blackouts and brownouts towards the end of the 1960s are likely to become even more severe over the next decade. The rationing of electricity, natural gas and even gasoline is a distinct possibility, and prices will certainly increase for the simple fact that environmental controls have to be paid for. It costs more to refine a lead-free gasoline and also requires a bigger crude-oil input; low-sulphur fuel costs more than one with a high sulphur content; it is more expensive to treat industrial effluent before it is discharged into rivers and lakes; devices to reduce carbon monoxide and nitrogen oxides from exhausts will put up the price of automobiles. The tragedy would be if, faced with the inevitable bill for these very ncessary controls, the government and public alike should swing as abruptly away from an awareness of conservation as they did towards it.

Ironically, the discovery of oil on Alaska's North Slope came not only at a time when it was most needed by the United States, faced with a growing energy crisis and a decline in production relative to demand from the traditional oil-producing states, but also when concern over conservation was reaching its peak. Blithely unaware of the fragility of the Arctic environment or that anyone cared very much, the oil companies moved in with one aim in mind, to explore for oil and, once having found it, to get it out as quickly and cheaply as possible. The first companies to drill on the North Slope in the mid-1960s operated only in winter, mainly for their own convenience and because time was not pressing. In so doing they caused the least harm to the Arctic ecology, although it is remarkable in view of what transpired later that they sought little advice from those already familiar with Arctic conditions, such as the US Naval Arctic Research Laboratory at Barrow or the University of Alaska. But the same cannot be said for some of the contractors employed by these and other major companies to carry out geophysical surveys in the area, working not only in winter but summer as well, when the Arctic environment is most vulnerable. Most of the early damage that marred the image of the industry as a whole was caused by seismic survey teams, trailing their bulldozer-drawn caravans backwards and forwards across the Slope and setting off explosions to record the returning sound waves and thus map the underground rock formations. It was not the explosions that caused damage but the movement of their vehicles and its effect on the underlying permafrost.

This is the description given to any ground beneath the surface that remains permanently frozen, even in summer. It is a peculiarity of the Arctic and Antarctic but is also found in some mountainous parts of temperate regions, accounting for over 20 per cent of the land area of the world. In one form or another permafrost covers 85 per cent of Alaska, including the whole of the northern area. It can be made up of frozen rock or veins of ice many feet thick, but more commonly in northern Alaska it is a mixture of silt and gravel and water, frozen concrete-hard up to depths of 2,000 feet. If it were not frozen, the ground would become a quagmire and ooze gradually into the Arctic Ocean, carried down by the rivers and streams flowing from the foothills of the Brooks mountains. The whole area could be eroded in a relatively short time. With temperatures in the short Arctic summer reaching as high as 50°F, this would be sufficient to melt the permafrost. The surface of the North Slope does in fact become marshy, laced with streams and thousands of lakes. But the permafrost underneath is protected from the sun's rays by a layer of vegetation, made up of grass, moss, lichens, as well as many varieties of wild flowers. This insulating blanket of tundra is only a few inches thick, but it is sufficient to prevent the permafrost from melting. If it is broken or cut away, however, then the permafrost is exposed and begins to thaw. Once started this becomes a self-feeding process that is almost impossible to reverse. The surface freezes again in winter, but the following summer the unprotected permafrost will continue to melt at an even faster rate, fed by water draining from the inland mountains. Ruts of no more than a few inches made by heavy tracked vehicles can rapidly become drainage channels and in a few years develop into chasms many feet deep, with little knowledge at present of where the erosion process might end. Even in winter the lowering of a bulldozer blade to plough too deeply through the frozen surface can have the same effect of cutting the tundra and exposing the permafrost to erosion when summer comes.

And that was precisely the result of some of the seismic operations as the crews hauled their heavy vehicles across the Slope. One fifty-mile seismic trench has eroded over fifty feet deep. There are trail marks all over the Slope which will remain possibly for ever to show where various teams have been. The ultimate symbol of the damage they caused, which perhaps more than anything else angered conservationists, was the carving of one seismic company's initials on the tundra in letters sixty feet high by a bored cat-skinner who had nothing better to do.

One man who knows as much as anyone about permafrost and its problems is Dr Max Brewer, director of the Naval Arctic Research Laboratory, who has lived at Barrow for over twenty years. The advice of his establishment was sought early on by some of the companies when they were considering designs for their North Slope camps—Arco's main base camp for instance is based on that of the Naval Laboratory itself—but it was not until criticism about the damage done to the tundra threatened to hold up oil operations that they thought to consult him about permafrost.

'The surface of the tundra is like a skin,' he explains. 'You break it—and it bleeds. In winter there are two skins, a layer of snow and under that a layer of ice. No harm is done if the top layer is scraped away by tracks, but if you cut through the bottom layer it is just the same as cutting through the tundra in summer. It is not too bad on flat ground, but it becomes really serious on tilting ground in the foothills where there is a downward slope. That is where the seismic crews have done the worst damage.'

The general response by the major oil companies to criticisms of environmental damage was that they could not be responsible for the operations of sub-contractors or of smaller companies who did not have a public image to be concerned about anyway. These were certainly the groups which caused the most harm, wanton in some cases and in others simply because it was the quickest and cheapest way to carry out the operation. It is true that big companies such as BP and Humble have a good record themselves over conservation, for one good reason that they are more vulnerable to criticism and have more to lose. But often the contractors were carrying out work on their behalf, under contracts which required this to be done as fast and as efficiently as possible and which certainly made no stipulations as to the methods employed.

Dr Brewer sees the cat-skinner as the villain of the piece, but that the companies which employ him must take responsibility for controlling his activities.

'The oil companies always had a rule that if a man could do a job, then leave it to him. To take a cat-skinner's blade away from him is like taking away his pants, but that is what has to be done. He wants to do the job like his father and grandfather before him—simply to plough straight on through with his blade stuck out in front. He should be told that if he doesn't do the work carefully, then he'll be fired. If that happened to a few of them, the word would soon get around and they'd behave themselves. The work, can be done with a minimum of damage by using proper Arctic engineering, but one of the problems is that a lot of information about the best way to go about it is not being disseminated by government agencies or by the oil companies themselves because of their competitive position.'

Dr Brewer is not one of those who want to ban the oil industry from the North Slope entirely.

'Man is part of the environment, which is something many ardent conservationists forget. The oil companies are speeding up the progress of the Eskimo. He wants things. In the past, he could not sustain communities of more than 300 because there was not that much game in the area. When I got here in 1950, Barrow already had 800 Eskimos and now there are over 2,000. The Eskimos used to be nomads living in tents. Now, while they've become more settled, new nomads have moved in—the oil companies. The basic trouble is their philosophy of impermanence.'

It is this attitude which caused another of the North Slope problems, that of pollution. Garbage strewn about old camp sites as the oilmen moved on to other areas, empty fuel drums dumped in lakes, seismic wire left on the ground to entangle passing caribou, as happened in a number of instances. Such examples of pollution are more damaging in the Arctic than anywhere else, for the region is like a huge freeze-box in which nothing ever rots. Waste materials will remain in their original state for many years, as can be seen on the Naval Petroleum Reserve from the junk left behind after the government programme of oil exploration nearly thirty years ago. The thrown-away oil-drum is one of the commonest sights all over the North Slope since oil is virtually the only fuel for heating and powering aircraft and land vehicles, and it can only practically be brought in in forty-five-gallon drums. Some of the oil companies have made efforts to clear their own drums away—BP, for example, found it could even make a small profit by flying them back in empty aircraft returning to Fairbanks and then selling them. But millions still remain, each one containing a small amount of oil. When this eventually leaks out, it cannot seep into the ground because of the permafrost and will instead drain into streams and lakes. A few drums would have little effect, but these tiny amounts of oil multiplied by millions pose a great hazard for the future.

If the early days of the oil industry's arrival brought problems, these were nothing to the impact of the oil boom in 1969 when the scramble by many new companies to get a stake in the Prudhoe Bay discovery before the September lease sale led for the first time to drilling throughout the year, including the vulnerable summer months.

'It was like a military invasion. Another Normandy landing, only better equipped and without so many scruples.'

That is how the one man living in the area before the discovery sees the arrival of the oil companies. Harmon Helmericks, once one of Alaska's most famous bush pilots who now makes a living as guide to hunters and fishermen, lives with his wife and three sons in what must be the most remote family home in the world, built in the late 1950s on the Colville River delta at the edge of the Arctic Ocean. Their modern two-storey, green-painted clapboard house, the kind one would find in an exclusive New York suburb, is the most incongruous sight in the wilderness of the North Slope. The furnishings inside—fitted carpets, piano in one corner and tropical fish tank in another, easy chairs and children's toys on the floor—are those of a successful business exective one might expect to see commuting to a big city every day. Only instead of a car parked in the drive, there is a Cessna light aircraft. When Martha Helmericks goes shopping, it is to stock up with provisions for the year. The nearest city is Fairbanks, over 400 miles to the south; their nearest neighbours are 150 miles away in Barrow. It is the ultimate in 'getting away from it all', the dream of so many living in crowded places.

'We used to have game all around here,' Helmericks recalls. 'Caribou, wolf, bear, wolverine. But we haven't seen them in any quantity for years. The seismic people used to hunt them from helicopters or buzz over them just for fun. There's not so much outright killing now, but the harrassment goes on. Take Pingo Island, a little way to the north. A herd of about thirty caribou used to live there all the year round. We never bothered them. Eskimos used to live there, as you could see from the graves they left. There were geese and all kinds of birds in summer. Well, the whole island was torn up by one company so that they could drill in summer. It could have been done in winter without too much damage but they had to find out quickly if there was any oil, which there probably wasn't anyway. Now the whole place has been destroyed.'

Helmericks is scathing about some of the oil companies' public relations advertising.

'They talk about turning the most desolate and forbidding place on earth into one of the most sought-after places. Sought after by who, I'd like to know, except the oil companies. Certainly not by the polar bears any more. They were run out on to the ice and have headed north as fast as they could. And as for being the most desolate place—the Eskimos didn't think so, and neither did the ducks and the geese and the caribou. They say the oil industry will provide jobs for the Eskimos in an area where there's never been any kind of work except hunting and fishing. What's wrong with hunting and fishing? I know many men who'd give anything for a life like that. Where we used to see twenty thousand caribou coming down through the mountain passes to graze on the North Slope in summer, now we're lucky if we see perhaps thirty. We're not against the oil companies but the way they've gone about things. Like the way they've blocked up the rivers so the fish can't get past.'

Apart from the disturbance to game and churning up the tundra, the effect on fish in the area is another controversial issue. When it was decided to continue drilling operations through the summer, a means had to be found to construct a network of roads and landing strips around Prudhoe Bay in order to move equipment over the marshy ground. The only material locally available was gravel from river-beds. Millions of tons were taken from rivers and streams in order to build these roads and to form five-foot gravel pads under the drilling rigs. This certainly prevented further erosion caused by driving tracked vehicles across the unprotected tundra but created another problem. For it is in these rivers that fish such as Arctic char and grayling come upstream every spring to their spawning grounds, sometimes eighty miles or more inland. The careless removal of gravel can block a river, or divert it from its normal path, or even upset the spawning grounds themselves. Androminous fish are very particular where they spawn, returning to the same places each year, and will not do so if the water has been muddied or the river-bed otherwise altered. Again the companies eventually came to appreciate the harm that indiscriminate removal of gravel could cause, but not before considerable damage was done.

To be fair, the blame cannot be laid entirely against the oil industry. The very nature of the competitive system imposed on them by state leasing compelled companies to carry out their drilling operations as fast as possible with little time available to worry about conservation. The state had its eye on a quick financial return, which it certainly obtained at the September lease sale, and actually encouraged the companies to drill in summer by allowing them to remove any quantities of gravel they required. Within the industry itself some companies made efforts to study and protect the environment—both Atlantic Richfield and BP employed ecologists to look into the problems and set in hand such research programmes as reseeding the tundra and studying the ability of Arctic waters to absorb and dispose waste. But again the competitive situation attracted many smaller companies which did not have these resources and whose only chance to make up for lost time, in the sense that they were not already operating on the North Slope, was to cut corners regardless of any damage that might be caused in the process. The more responsible companies could not themselves impose controls on the operational methods of smaller groups. Only the state and federal authorities could have done that, on whose lands the companies were operating. And neither the state nor the federal government had an Arctic policy to cope with the invasion of several thousand men and their equipment on to the North Slope. The State Department of Fish and Game, for instance, employed only one law enforcement officer on the entire North Slope. (There are only eight in the whole 400,000 square miles of northern Alaska.) His almost impossible job was to protect game such as Arctic fox, wolf and caribou from illegal hunting without a licence, which costs $100 for non-residents and three dollars for residents who must have lived at least a year in Alaska. Most of the oil-camps had rules against the possession of firearms, but the number of oilmen charged with hunting without a licence and the dwindling amount of game in the area argue against their wide enforcement. One officer found he actually had to charge an oil company ecologist for shooting two bears and leaving them without skinning them or using their meat, a cardinal sin in Alaska where the ideal at any rate is that a man should only kill game for its meat. But even this is negated by the fact that in large areas of Alaska a state government bounty is still paid for killing wolves, though no one knows how many or how few wolves might be left. With a wolf being worth nearly $200, including the bounty and the sale of its skin for the parkas that are such essential wear in winter, most are killed from aircraft by professional hunters, in some cases machine-gunning whole packs. Again the polar bear population in the Arctic is thought to be diminishing and yet Alaska is one of the few countries within the Arctic Circle that permits their hunting by others than Eskimos, who partly rely on them for their subsistence by selling the hides or using them for clothing; the meat cannot be eaten by humans. Russia allows no hunting of polar bear at all, Canada allows less than 400 to be taken annually by natives only. The annual 'harvest' allowed in Aalaska is 300, of which just seventy-five may be taken by Eskimos. The remainder are mostly killed by non-resident trophy hunters who search for them over the ice in light aircraft. With such an expedition costing around $3,000, including the licence and fee to a guide, this brings an income into the state which, like the wolf bounty, has made it politically impossible so far to ban such hunting although there are many Alaskans who would like to do so. It is one of the anomalies about conservation that some of the most dedicated preservationists are the most ardent supporters of hunting, not just for meat but for trophies as well, their main desire to preserve the wilderness primarily being for their own sporting pleasure.

But whatever the reason, by the time the conservation issue had become a potent force in 1969, the oil companies were already firmly entrenched on the North Slope. Oil had been discovered and the state had received its share in the bonanza by the $900 million lease sale. Those who were determined to fight the oil industry on conservation or political grounds could do little to affect operations on the North Slope itself. Until 1970 no visitors were even allowed there, unless by specific invitation, and in any case the oil companies were beginning to put their own house in order, removing some of the debris and forming, for instance, a multi-company Environmental Committee to consider such problems as organic and solid waste disposal, oil spills and pollution control, fog problems resulting from flaring off natural gas before a pipeline could be built to flow it to the consuming areas, tundra preservation, and the care of fish and wildlife. But where they were most vulnerable to attack was their need to build the trans-Alaska crude-oil pipeline, without which the oil would have to remain uselessly in the ground. It was against this project that the conservation groups directed their main criticism, centring on three fundamental points—permafrost, earthquakes, and caribou.

When the seven member companies of the Trans-Alaska Pipeline System, led by Atlantic Richfield, British Petroleum and Humble Oil, began in 1969 seriously to study the engineering of the line and how it could best be built, they knew they faced formidable problems. They knew, of course, where it had to start at Prudhoe Bay as a gathering centre for the many smaller lines bringing oil from production wells spaced out over the field. And early on they had decided on Valdez, a deep-water port in Prince William Sound, as the most suitable terminal location where the oil would be collected in large storage tanks and then transferred to tankers for shipment to the West Coast refineries. But between these two points the pipeline had to cross some of the most difficult and inaccessible terrain in the world, and its exact route could not be calculated until more detailed surveys had been carried out. Broadly, it had to cross about 100 miles of the North Slope itself, then another 100 miles over the mountain passes and valleys of the Brooks Range, through 350 miles of forested hills and rivers of central Alaska, then 100 miles of mountainous country over the Alaska Range, down through the swampy Copper River Valley for another 100 miles and up again over the rugged Chugach Mountains and along the Keystone Canyon for the last fifty miles to the coast, the final section involving the most difficult mountain pipeline construction ever attempted. With some of the mountain passes rising to an altitude of 4,700 feet, five pumping stations would be required at intervals along the route to keep the oil flowing at an initial capacity of 500,000 barrels a day; seven more would be needed to provide for the eventual two million barrels a day planned. The extreme cold over some of the sections posed special technical problems. Even though the oil would leave the wellheads at a temperature of about 160°F it would soon cool and become too sluggish to move. However, the pumping stations themselves would provide heat to keep the oil freely running and were so designed to maintain the temperature of the oil and the pipe at about 100°F throughout the entire length.

But a hot oil pipeline, although the only feasible solution was obviously going to create even bigger problems due to the presence of permafrost over most of the proposed route. The most practical way to lay any pipeline is to bury it several feet in the ground where it is protected from the weather and any accidental or even intentional damage and where it is also less of an eyesore. Where the permafrost consisted of frozen gravel or rock there would be no problem in burying it, but in those places where it was mostly frozen silt and ice, such as the North Slope, the heat of the pipe would cause this to thaw with the resulting kind of erosion that had already been experienced during the early oil operations. As the ground melted and subsided, so would the pipeline, sagging to the point where, with a weight of 1,000 lb a foot when filled with oil, it would break and the oil spill out. The only answer in the extremely delicate ice-rich areas was to build the line above ground on piles or gravel support pads, ensuring that it was high enough and sufficiently insulated so as not to heat the ground beneath. This would be a more costly operation and, as far as the engineers were concerned, should be avoided unless there was no other way.

Very little was known about the exact nature of the permafrost in Alaska, or which areas had the highest ice content, when TAPS engineers began working on the project. In the spring of 1969 they drilled 300 bore-holes to take core samples along the proposed route, discovering in some areas as far south as the Copper River Valley that while one section of ground might contain a wedge of almost solid ice many feet thick, only a matter of yards away to one side the permafrost might be made up mostly of gravel and rock in which it would be safe to bury the pipeline. It was found that the problem on the North Slope could largely be overcome by routing the line along the banks of the Sagavanirktok River where the gravel deposits made it possible for the line to be buried. This meant taking the line over the Brooks Range by a 4,700-foot unnamed pass into the Dietrich River Valley rather than the 2,400-foot Anaktuvuk Pass, presenting more severe hydraulic conditions. But there were other reasons also to do with caribou migration that made TAPS decide on the Sagavanirktok route. From all the various tests and surveys carried out in 1969, TAPS reckoned that only about fifty miles of pipeline would have to be built above ground in the really difficult places and that the rest could be buried. The exact route would have to be selected as the work progressed, zig-zagging to avoid high-ice areas, but they felt this could most economically be done while the pipeline was being laid. The other major technical problems were crossing the Yukon River, which had never been spanned before except by winter ice when the river freezes to at least 16 feet down, and earthquakes and earth tremors which are highly prevalent in Alaska, as witnessed by the 1964 disaster in Anchorage and Valdez. In the case of the Yukon it was thought at first that an overhead bridge would have to be built, but as a result of drilling bore-holes in various places when the river was frozen over, an area of solid bedrock near Hess Creek was discovered in which a trench could be blasted and the pipeline laid under the river. The possibility of earthquakes is more problematical, especially in the Alaska Range where the major Denali Fault has shown signs of recent activity. Nothing could prevent the line from breaking in the event of a severe earthquake, but special precautions were built into the design to provide for a monitoring system to detect any creeping movement in potential earthquake areas and valves to cut off the flow of oil should a breakage occur.

One essential requirement was for a road along the entire route of the pipeline, both to bring in the necessary construction equipment and to allow maintenance to be carried out afterwards. Roads already existed over most of the route from the south coast to Fairbanks and up as far as Livengood, but from there northwards it meant building an entirely new road; it would in fact be the first road ever built connecting Fairbanks with the Arctic coast. A trail had been bulldozed through in the winter of 1968–9 to carry oil equipment up to the North Slope but this ice road, as it was called, had proved something of an environmental disaster, causing severe erosion of the permafrost in some areas and land slides where it wound along the sides of hills. However, by using gravel from rivers and hill-tops, a permanent all-weather road could be built. Encouraged by the Alaska state authorities, who wanted to see the oil moving out as quickly as possible and thus obtain a revenue from royalties, a start was made by repairing the haul road from Livengood to the Yukon. The steel pipe was ordered and shipped in from Japan. Construction work was put out to tender and fifty million dollars' worth of special road-building machinery assembled. The oil companies were confident that by completing the road to Prudhoe Bay by the end of 1970 and building the pipeline in two segments, starting the 430-mile southern section from Valdez to the Yukon in March 1970 and the 368 miles from the Yukon to Prudhoe Bay at the end of the year when the road was built, they could keep to their original schedule for beginning to pump the oil out by 1972.

But there was a hitch. Over 90 per cent of the route lay across federal land, which meant that the Department of the Interior had to give its permission for the necessary 100-foot right-of-way for the pipeline and 200-foot right-of-way for the road from the Yukon to Prudhoe Bay. Had this been state land, authority would have been given without question. But the federal land freeze imposed until a settlement of native claims had been reached made it impossible for the state to set aside the right-of-way as part of its land selection and assign it to the oil companies. The total area of the route was less than twenty square miles out of Alaska's 586,000 square miles. As a relative comparison it was narrower in width than the green line painted down Fifth Avenue on St Patrick's Day is to the whole of New York City. But cutting right across the centre of Alaska, for the first time connecting north and south, it had an emotional impact far greater than the actual amount of land involved, especially when seen as a black line across a small scale map of the state. The Sierra Club, one of the largest and oldest of the conservation organisations in the United States which had been aggressively flexing its muscles since mounting public concern over pollution had given it a new power, described the proposed pipeline as 'an environmental horror that would put an 800-mile scar from Prudhoe Bay to Valdez'. Less emotively, the US Geological Survey expressed dissatisfaction with what it considered too meagre data put forward by TAPS in trying to show that the line could be built without great hazard. In the wake of the pollution caused by the *Torrey Canyon* and Santa Barbara disasters, the possibility of oil gushing out from a break in the line and contaminating the wilderness of the last frontier was enough to make most people shudder and bring the strongest protests from conservation groups. Because of the permafrost problem, officials of the Geological Survey reckoned that 90 per cent of the line might have to be built above ground and only 10 per cent buried, the opposite of what TAPS had intended.

This raised another problem, this time to do with caribou. These large North American deer are the most hunted of any animal in Alaska, a main source of subsistence for Eskimos, Indians and the many white families who go up into the hills to find and kill their own meat. There are over 600,000 caribou throughout Alaska of which some 400,000 exist in two large northern herds, known as the Arctic and Porcupine. Most of these animals spend the winter south of the Brooks Range and then in spring migrate through mountain passes to the summer grazing grounds of the North Slope, returning the same way with the onset of winter. One of their main routes is through the Anaktuvuk Pass, where at certain times of the year the snow is black with thousands of slow-moving caribou and where generations of Eskimos have lain in wait to hunt them. There have been temporary hunting settlements here since prehistory and a permanent village of Anaktuvuk since early in this century, whose inhabitants are among the few Eskimos living their traditional way of life with little contact so far with the outside world. The village is built at the bottom of the pass some sixty miles from the tree-line. This means that wood for building and fuel must be hauled that distance down from the hills—but it is easier than having to haul a dead caribou the same distance if the village were built higher up. In some instances, where caribou have made a detour some way off, attempts are made to shoot them in one leg and force them to limp to the village before they are killed, in order not to have to carry their heavy weight; a cruel practice to outsiders perhaps, but values change when one's existence might depend on it. Several years ago, when for some inexplicable reason the caribou took another route through the Brooks Range and completely avoided the Anaktuvuk Pass, the Eskimos of Anaktuvuk faced starvation and were only saved as a result of the US Air Force dropping food supplies.

The most critical time for the caribou herds is during their spring migration when calves are being carried. The calving grounds are in specific areas to which the caribou return each year—near the head waters of the Colville River for the Arctic herd, and in the foothills of the Romanzof Mountains and the British Mountains over the border in Canada for the easterly Porcupine herd. Timing is essential. The caribou must cross the Brooks Range while the snow is still firm enough to support them and reach the calving grounds at the right moment early in June to drop their calves. A delay of only a few days would cause the calves to be dropped in less suitable places, resulting in a very high mortality rate. Any large-scale construction of the road and pipeline might cause the caribou to change their route to by-pass it and thus lead to such a delay. As a result of caribou studies carried out by Bryan Sage, BP's ecologist attached to TAPS, who in 1969 and 1970 covered the entire route of the pipeline on foot, the work schedule can be arranged to avoid construction during these critical times. But a bigger problem is the effect that a permanent above-ground pipeline might have on caribou migration—whether the animals would skirt around it in places where it was buried, or step over it if special crossing structures were erected. This is still largely an unknown factor.

The disagreement between TAPS and the Geological Survey as to the length of pipeline that could safely be buried led to many more studies by both groups in 1970, including 2,000 more borings along the route, less than twenty-five feet apart in the most critical places, and tests and research into every aspect of environmental hazard. For what TAPS now had to show beyond any doubt, under the terms of the National Environmental Policy Act which became law on January 1, 1970, was the precise impact on the environment that would be caused by building the pipeline. It was no good assuming or just hoping that it could be built without much environmental damage; this had to be proved to the satisfaction of the newly formed Council of Environmental Quality. It was as a result of this Act that TAPS suffered the most damaging blow of all to its hopes of making an early start on the pipeline. On April 13 Judge George L. Hart, Jr, the same Federal District Court judge in Washington DC who, earlier in the month, had granted the Indians of Stevens Village a preliminary injunction against the Interior Department giving a right-of-way permit for the proposed pipeline to cross a twenty-one-mile stretch of village land, also granted an injunction at the request of three conservation groups to prevent the Interior Department issuing a right-of-way permit for the road from the Yukon River to Prudhoe Bay until the issue could be tried on its merits, which meant conforming to the new environmental laws. Since the oil companies could not begin the northern section of the pipeline until the road was built, and did not want to chance building the road until the exact pipeline route was settled and permission given for its construction, this effectively halted any further work on the project. There were detailed arguments about the width of the right-of-way required, and whether or not this came within the limits set by the 1920 Mineral Leasing Act. The point was also made that, once the pipeline had been built, the road would be handed over to the state of Alaska as a public highway. But this did not alter the outcome, with Judge Hart stating: 'It's as plain as day that the purpose of the road is to build this pipeline', and insisting that they were both part of the same enterprise.

The three groups who brought the complaint, the Wilderness Society, Friends of the Earth and the Environmental Defence Fund, were all national conservation organisations with headquarters in Washington and by no means represented similar groups in Alaska. Neither the Sierra Club, which had a chapter in Alaska, nor the Alaska Conservation Society, which in the early 1960s had been one of the proposers of the Arctic Wildlife Range and successfully fought the Rampart Dam scheme for damming the Yukon and flooding nearly 11,000 square miles of country, had joined in the suit. The Wilderness Society had no chapter in Alaska and Friends of the Earth was formed by a former executive director of the Sierra Club, David Brower, who had resigned over a major controversy to do with an aggressive and expensive publications policy. Some Alaskans supported the injunction, but many more felt it was an example of the kind of outside interference in their affairs to which Alaska had been subjected for so much of its history. Local industry and business groups, facing financial loss after gearing up to undertake the required construction work, were naturally incensed. There were warnings by officials of the state government that Alaska might be bankrupt by 1976 if oil royalties had not started coming in by then. Even among ardent conservationists in Alaska there were differing opinions. Some wanted the pipeline built for its economic advantages but not the road, fearing that it would lead to further development, further roads, and gradually open up the whole of the northern region whereas they wanted it kept as an inaccessible wilderness. Others were in favour of such roads because they would allow hunters and campers greater access to the countryside but were against the pipeline as a pollution hazard. Others were more concerned with the threat of spills along the south-eastern coast as a result of tanker shipments from Valdez. Some were concerned with preserving wildlife for its own sake and as a means of subsistence for natives; others were more concerned with preserving animals so that they could go out and hunt them for themselves. It was small wonder that the whole question of the trans-Alaska pipeline should become bogged down in a maze of contradictory opinions, led by small but aggressively vocal groups who found it a handy weapon with which to hammer the oil industry.

The attitude of TAPS management itself had not helped matters. At the outset, apparently unaware of the growing emotional appeal for conservation or relying on the strength of the oil industry lobby in Washington, the company did little to explain publicly the work it had done on the project and which seemed, to its own satisfaction at any rate, to make the pipeline feasible. It maintained an obstinate silence, preferring to deal directly with government officials which past experience had shown was the best way of getting projects approved with the minimum amount of trouble. This gave the impression that TAPS had done even less of its homework than was the case, and played into the hands of the conservationists who were determined to fight the pipeline all the way. The company also made a miscalculation politically. As Governor of Alaska when the oil strikes were made, Walter J. Hickel had hardly shown himself to be on the side of the angels as far as the conservationists were concerned. He believed in exploiting the state's natural resources, and he it was who gave permission for the ice road to be driven through from the Yukon up to the North Slope in order to move in oil drilling equipment. It was even dubbed, ironically, the 'Hickel Highway'. When, early in 1969, he was appointed Secretary of the Interior in the Nixon Administration, the key position for any decision on a permit for the road and pipeline, the oil industry felt it had a friend in court. It went ahead with ordering the pipe and putting out tenders, confident that the necessary permits would come through quickly. But, once in power, Hickel surprisingly became one of those most dedicated to environment protection, outdoing the moderate conservationists in government as if seeking to disprove his previous image as an exploiter. Not only did he delay any start on the pipeline until the effects on permafrost could be adequately studied, but he also stopped all drilling in the Santa Barbara Channel because of the danger of further oil spills and cracked down on the oil industry in many other respects. So far did he take such measures, shrewdly sensing the anti-company attitude of America's young, that in November 1970 he was summarily dismissed from the Cabinet by President Nixon. Only a short while later, during public hearings on the pipe-line, Hickel came out in support of it, stating that North Slope oil was essential to the nation's economy in view of dwindling reserves elsewhere, and that if nature had to put oil somewhere it was better on the remote North Slope of Alaska than in the Tetons or the Sierras or Cape Cod. But the months of delay and controversy had given the militant conservationists time to whip up even more public feeling against the pipeline. The new Secretary of the Interior, Rogers C. B. Morton, found himself with a political hot potato in his hands that made it impossible for him to give his quick approval without also giving the impression that Hickel's dismissal and his own appointment were the result of pressures from the oil lobby.

Meanwhile, the oil companies had been amassing a vast amount of data for the environmental impact statement required by the National Environmental Policy Act, which would then be subject to comment by federal and state agencies and the public to form the basis for a final settlement. At the same time they found it convenient to change the name of TAPS to the Alyeska Pipeline Service Company under the presidency of Edward L. Patton from Humble Oil, with engineering offices in Houston and a field office in Anchorage under BP's David Henderson. This was done both to streamline the organisation and iron out some of the problems of inter-company communication; behind the scenes under the former TAPS management there had been considerable disagreement between some of the partners as to the best way of getting the project going. From the start BP, more accusomed than the American companies to dealing with sensitive political situations in foreign countries such as the Middle East, had favoured a more open approach to both the government and public to show that they were aware of the conservation problem and were taking steps to resolve it. Some of the American partners, on the other hand, believed that the power of the oil lobby could be relied on to see the project through without having to divulge much either publicly or privately about the way it would be built—a mistaken belief as it turned out, in view of the changed climate of political opinion.

By the time the draft environmental statement was produced early in 1971 and presented for comment by the Interior Department, Alyeska and the US Geological Survey had compromised on an agreement that at least 52 per cent of the pipeline could safely be buried, a far cry from the 90 per cent and 10 per cent respectively that the two groups had predicted eighteen months earlier. The Geological Survey found the remaining 48 per cent to be questionable, pending further soil studies, but in any case Alyeska would elevate the line where thawing of ice-rich permafrost could be a problem. A point not always appreciated by those criticising the oil companies was that it was just as much in their interest to build a safe pipeline that would not break and waste the oil by spillages. The impression given by some was that the companies almost wantonly sought to cause pollution, such was the emotion generated by the controversy. On the other hand arguments that the cost of the line had now doubled because of the delay, and might increase to two billion dollars if a start was not made quickly, were equally specious. The longer the oil remained in the ground the more valuable it would become, as evidenced by the 1970 price increase of twenty-five cents a barrel for US domestic crude oil. And if the pipe itself was not already purchased and had to be ordered in 1972, it would cost something like 40 per cent more.

It seemed that agreement on the pipeline was on the point of being reached. The Interior Department and other federal agencies supported it, especially in view of the threatened energy crisis and the stringent restrictions imposed to ensure an absolute minimum of danger to the environment. But, at public hearings held in Washington and Anchorage, what received the most publicity were a few phrases from the lengthy statement that suggested some environmental damage would inevitably follow the construction of the pipeline. This was patently obvious. Any kind of incursion into virgin wilderness would cause some damage—the much vaunted country-loving campers and hunters themselves carelessly caused many of the fires that in 1969 destroyed 4·3 million acres of forest in Alaska. There were times when Anchorage itself was nearly obliterated by smoke. Any idea of Alaska as a pristine haven for the pollution-conscious is discounted by the beer cans and litter strewn along its roads and the garbage left in hillside camps, no worse than in other parts of America but certainly no better. But the pipeline controversy had now left the bounds of reason. It had become a symbol not just of the need for conservation but of political attack on big companies, and particularly the oil industry. Genuine conservationists found themselves in strange company when the extremist Weathermen organisation, for instance, responsible for some of the most vicious bomb outrages in America, should see fit to include pollution of the environment in their hazy condemnation of big corporations.

Another red herring was raised when a Canadian parliamentarian, testifying at the hearings, suggested that tanker traffic between Valdez and the West Coast might prove to be a hazard to the Canadian coastline in the event of collision. This led to another outcry and demands for reconsideration of a pipeline route across Canada to Chicago that would not cross earthquake zones or involve the use of large oil tankers. This suddenly seemed to be the magic answer since it was tankers that were now bad news after a number of well-publicised collisions, forgetting somehow that such a line would have to cross the North Slope Wildlife Range, that the permafrost and wildlife problems were just the same in Canada, and that the line would be twice as long as that across

Alaska. But, bowing to the weight of conservationist opinion, the Interior Department promised to look into this possibility, in the meantime holding up a decision on the trans-Alaska pipeline which meant further delay. A revised and 'final' statement, an essential prerequisite to removing the conservationists' injunction, was published in 1972—the year in which the pipeline originally should have been completed. Whatever happens it cannot now be completed before 1975, by which time its cost will have escalated to around $3 billion. Meanwhile the rust gathers on the stacks of steel pipe, the trenching and roadbuilding machines wait motionless, drilling on the North Slope limps nearly to a halt. The oil industry, with over $1·5 billion in invested capital tied up in Alaska for nearly three years, is losing $2 million a day in both loss of income and interest on this capital. And the state of Alaska, which is also failing to receive the $750,000 a day in revenues it expected from oil production, heads back towards bankruptcy. One oil company has already decided that too much is enough. Marathon Oil has withdrawn entirely from northern Alaska, even though it meant writing off $15 million in North Slope leases. It may not be the last company to do so. In understandable frustration the Governor of Alaska, William Egan, has proposed that the state should take over the pipeline by a kind of compulsory purchase—which even if practical would cause yet further delays. But ideas are the one commodity not in short supply. The pipeline company has been inundated with hundreds of transportation schemes from home inventors and big corporations alike, including the use of aircraft, airships, submarines, trains, and even a huge diameter pipeline through which tractors would pull strings of oil-wagons—as long as men could be found who would not feel claustrophobic about driving through 800 miles of tunnel! In this and many other ways the whole project has entered a world of fantasy—in fact, a kind of oil in wonderland.

17

Oil in Wonderland

On the desk of a senior oil company executive in London, a man usually concerned with high-level government negotiations and whose daily work brings him into contact with Cabinet ministers and Middle East sheikhs, there is a detailed report on the breeding habits of the Lapland longspur. It might just as well be about brant geese or eider duck or any other rare waterfowl of the North Slope. It is given all the concentration that might be devoted to a new oil agreement involving millions of dollars.

'All I seem to have done this past year is study Alaskan wildlife,' he says ruefully. 'What do you think? Will caribou jump over the pipeline or take the long way round?'

An engineer in Anchorage is trying to work out just that. He is designing gravel ramps to go over the pipeline in those places where it is built six feet above the ground. It is hoped that migrating caribou will use these ramps to cross over the obstacle.

'This is where the birds and bunny people have got us,' he says, meaning the conservationists. 'We're bending over backwards to meet their objections and still they're not satisfied.'

A burly Texan construction worker driving up to the Yukon stops his truck, picks up an empty beer-can and stows it away with his belongings to be discarded later.

'It was probably thrown away by a hunter,' he explains. 'But we'd only get the blame for it. Better make sure. We got our orders. Anyone throwing trash away—out. Fired on the spot. You know something? I reckon this is the cleanest road in the States.'

He's probably right. Smarting under attack by politicians and conservationists alike, the oil companies are leaving nothing to chance. They want to prove their good housekeeping and care of the wilderness so that they can get on with the job they came to do. Committees of tough oilmen will be found arguing passionately about whether the removal of gravel from a particular section of river will affect the spawning grounds of grayling, where a few years ago one method of fishing by military personnel on leave in Alaska was to throw grenades into a lake and scoop out as many of the dead fish as they wanted. The migration of caribou and the lambing season for dall sheep are as important matters for discussion as the size of drill pipe and the weight of drilling mud, but wealthy Texans and German industrialists still continue to hunt the diminishing polar bear over the Arctic ice. In some of the oil company camps on the North Slope sewerage is carefully packed into containers and flown back to Fairbanks for disposal—where, because of a lack of proper facilities anyway, it is usually dumped into the river. In the summer of 1971 a group of scientists studying bird life on the North Slope reported a serious reduction in the number of eggs laid by sandpipers that year in the nests they had examined. This threatened to blow up into another major ecological row until it was discovered that a crafty fox had worked out that by following the scientists on their tours of inspection, they would lead him to the best nesting areas from where he would take the eggs after they had left. Birdwatchers now scatter mothballs after them to put foxes off the scent. There is a kind of unreality about it all. It has even been seriously suggested that because the pipeline will warm the ground in those areas where it is buried, this will so improve the quality of the re-seeded grass that the caribou might be unduly attracted to it and destroy the vegetation by over-grazing, in which case the pipeline itself would have to be protected from wildlife by fencing. Such examples of where the conservation drive has taken the oil industry cause a certain wry humour. But there is also some bitterness, and not just on the part of those seeking to protect the environment.

'They got us to spend several billion dollars in Alaska,' says one disgruntled oilman. 'Then, when we'd put up the money, they turned round and said we couldn't take the oil out. How dishonest can you get? If some country in the Middle East had done that, the government would probably have sent a gunboat in.'

In some respects, the oil industry has only itself to blame. The damage that was done in the early days on the North Slope was thoughtless more than wanton, but it was no excuse to say the contractors were mostly reponsible for they were working on behalf of oil companies. In fact the overall degree of damage was less than would have been caused by any other industry. Oil drilling does not even begin to compare with the scarring of the landscape caused by open-cast mining or pollution from manufacturing plants and factories. But it was the attitude of some of the oil companies that was resented as much as anything else. The image projected in advertising and publicity of tough oilmen battling against nature and taming the wilderness might have been acceptable in the past as a human pioneering endeavour to be proud of, but it was at odds with the new enlightened realisation of the need for conservation. Then again, the pipeline project was embarked on and regarded as a *fait accompli* before the oil companies under the leadership of Humble Oil were absolutely certain of how it could be built. A greater consideration of the permafrost and earthquake problems had been given to it than people thought, but the TAPS organisation saw fit to keep this very much to itself. There was little attempt to inform the public, or get together with the government authorities to discuss the matter. The typical response of the American companies was: 'Don't go near Washington, don't talk about the problems because it will only complicate matters. We can fix it.' The only trouble was that the oil lobby could no longer fix such deals behind closed doors. When the issue became a matter of public debate and controversy, fed by the opposition of conservation groups, it seemed that the oil companies had more to hide and had paid less attention to environmental problems than actually was the case. More than anything else the project was a failure in public relations. It was even rumoured that Humble's parent company, Standard Oil of New Jersey, was intentionally dragging its feet over the pipeline project because it did not at that time need North Slope oil, having sufficient supplies for its own needs from other domestic sources in Louisiana and elsewhere, and was not about to help the competition that would result from the new BP–Sinclair–Sohio giant marketing Alaskan oil in the United States. Whatever the truth of this, it is significant that it should have been widely believed in Alaska.

Some companies such as Standard Oil of California, BP and Atlantic Richfield paid close attention to the need for public relations and were as helpful and open about their operations as they could be. Other companies seemed to think that by saying as little as possible the controversy would die down as a passing phase. This certainly appeared to be the attitude of the London public relations consultants of one major American company which, ironically enough, had one of the best records of any company for its care of the environment while drilling on the North Slope.

In middle-class Alaska the two main subjects of discussion are state politics and oil. They are often related.

'Our politicians are too stupid even to be corrupt. They're in the hands of the oil lobby, all the way along the line.'

The speaker is an Anchorage businessman who sold his North Slope leases a few weeks before the Prudhoe Bay discovery was made and saw a million dollars slip through his fingers. He has been bitter about it ever since. There are in fact over 100 registered lobbyists in Juneau, many of them representing oil companies. They deal with what must be one of the most informal state legislatures in the country, where one Joint Session of the House and Senate was opened by a travelling pop group singing 'Up with people' and other protest songs. (As one member commented, 'This is the first time a Joint Session has been turned into a joint.') But the way things have turned out, the much vilified oil lobby has not been all that powerful or successful.

'We just haven't got the right calibre of guy in politics these days,' says another Anchorite. 'I guess everyone's too busy making money.'

Names are considered and discarded in the game of 'shoot the politician' like the snow-flakes melting on the window-panes. Party loyalty doesn't mean too much here. In a state where the population is so small that one has the impression everyone knows everyone else, it is personalities that count. When it really comes down to serious matters, the most heated discussion is whether the traffic lights in a new suburb should hang over the centre of the street as they do in down-town Anchorage. Politics are indeed local in this land that is more than half the size of Europe. And in the prevalent conservation-minded atmosphere, even the question of unsightly traffic lights becomes one of protecting the environment. Like oil.

'It's a fight, but we're going to protect our wilderness.'

The woman who speaks so passionately about conservation is a member of the Wilderness Society. She believes the oil companies should be sent packing and the wilderness left as it's always been. Whether anyone goes there doesn't matter. It is enough to know it is there. There is no arguing with the ardent conservationist that a man-made countryside of farmland and meadow can be beautiful too, or even cities where they have been designed by artists and not planners. It is only the wilderness that counts. And human beings don't come into the scheme of such things. Anything touched by man is contaminated.

But while the argument about conservation rages, waiting patiently for a just settlement of their claims are the Eskimos and Indians of Alaska. This controversy at least is something for which the oil companies cannot be blamed; in fact their operations can directly help the natives in providing employment if they wish it, and indirectly by providing the means to make a financial settlement possible. Without denying the validity or sincerity of their protest about protecting the environment, most conservationists are strangely silent on what to some is the greatest of all the injustices done to Alaska or her people. The Alaskan natives suffer from the curious paradox that while the United States has always been a colonial power in every sense of entering and dominating lands occupied by other races, she has preferred to pretend that she is not, often condemning other colonial nations for just that reason. This meant that the United States never had a colonial policy to cope with subjugated people such as the Eskimos and Indians of Alaska other than the myth of equal opportunity as American citizens. The system offered no chance for other races except to assimilate the American culture, which meant giving up their own. Where the consciences of other colonial nations were awakened to the point of returning lands to their original inhabitants, or of being compelled to do so by the weight of US and world opinion or the force of national aspirations, the most that American conscience could do was to relieve poverty by welfare handouts, serving only to reduce native pride and culture even further. Even now there is strong opposition in Alaska to the native claims Bill; even now many white Alaskans regard Alaska solely as their country and resent the charity given to layabout natives. It is one of the ironies of the situation that those Alaskans most in favour of the oil pipeline being built, with its resulting economic advantages, are also those most against the terms of the proposed settlement of native land claims. And yet, because of the federal land freeze, the first cannot be approved without the second. Like the conservation issue, there is a peculiarly Alaskan mixture of idealistic and selfish motives.

Alaska is unique amongst American states. As America's last frontier it is grappling with problems to do with native rights and conservation that were simply ignored in the days when other states were being developed by ruthless exploitation and determined pioneering. It is the only state in which the vast majority of its land is still federally owned, and yet because of its geographic position it is dealing increasingly as a nation with other countries such as Japan, Russia, the Pacific islands and Europe, with some Alaskans foreseeing a time when it could have its own foreign affairs representatives abroad and separate treaties covering agreements over fisheries and shipping. It is a wilderness still, and yet it possesses the greatest mineral and natural resources of any part of the United States which will be demanded with increasing urgency to maintain America's economic growth.

Oil and natural gas are, of course, the immediate cases in point. One way or another, both will be moving out of northern Alaska by the end of the 1970s. Even such conservationists as Bob Weedon, representative in Alaska of the Sierra Club and other groups, admit this; the only argument is the best way to do it.

'We are making small gains in a battle that is largely being lost,' he says. 'The important thing is to move slowly and study all the alternative methods to find the least harmful way of getting the oil out. I know the oil company ecologists have been making studies, but they are largely superficial. For someone to come up here in one summer and prepare a report on the pipeline route, never having been here before—it's a bit far-fetched that it could be really useful. And when it comes to offshore drilling in Bristol Bay, which has the biggest red salmon fishing industry in the world, it makes me shudder. With the currents there, it would only take a minor oil spill to ruin the whole region. Offshore drilling techniques haven't been sufficiently developed, and yet one company is drilling in that area right now.'

For it is not just the North Slope but most of Alaska, with its large number of sedimentary basins and potential oil-bearing areas, that is being speculatively eyed by the oil industry. One of the problems is that only recently, after a century of abundant supplies of cheap fuel which made possible America's economic prosperity, has the country become aware that it faces an energy crisis. It always seemed a situation that could never happen, consequently the United States has never had an overall national energy policy. Individual fuel industries were run with varying degrees of federal and state control, some strictly as in the case of natural gas, others allowed to compete freely in terms of price such as oil and coal. Scattered throughout Washington are over sixty separate governmental agencies and congressional committees that in some way influence or establish policies for oil, gas, coal, nuclear power and hydropower. Seldom are plans co-ordinated on an overall energy basis, although moves are currently being made in that direction.

Meanwhile, the United States now shares a world problem. In the past ten years the consumption of oil has grown 50 per cent in the US, 250 per cent in Western Europe and nearly 600 per cent in Japan, and is increasing all the time. Over the next twenty years the free world will consume about 500,000 million barrels of oil and 750 million million cubic feet of natural gas, which amounts to more than 80 per cent of the present proved free world reserves of oil and seventy-five per cent of natural gas reserves. Even to find this amount and maintain the present twenty-year reserves-to-production ratio would pose an immense problem for the oil industry, more than half the total discoveries made since the industry began. But by 1990 demand will have increased to about four times that of today. Those amounts several times over will have to be found to maintain a safe inventory until the end of the century. There is still undoubtedly a great amount of oil and gas to be discovered, expecially in offshore regions, but the supply is not inexhaustible. Unlike most other mineral resources such as copper and iron which are extracted from the ground to be used by mankind but still basically remain in one form or another for further use if necessary, petroleum is a wasting asset. The world oil industry reached its maturity in the early 1950s when the Middle East discoveries and pricing agreements paved the way for an international oil boom that increasingly displaced coal and has made the world dependent on oil as its main energy source. Reserves will inevitably decline, however—the United States is symptomatic of this—and oil as an energy industry as it is known today will have died by about the year 2040. Oil and natural gas will still be used, processed synthetically from the world's vast reserves of coal, shale and tar sands, but their place as energy fuels will be taken over by new forms such a nuclear and solar power and fuel cells which will also have the advantage of being cleaner. Following the current trend petroleum will increasingly become a chemical industry, providing the raw materials for plastics, fertilisers and chemicals.

Until the new energy technology is available, and it cannot begin to have an impact for twenty years or so, the world remains dependent on the discovery and production of massive quantities of fossil fuels. The threat to the environment during the intervening period is very real. From the production end there will inevitably be oil leakages at sea as the industry moves from land to offshore, increased by the greater amounts of oil being moved about the world in tankers. From the point of view of consumption, the burning of ever-increasing quantities of oil will add to the already serious air pollution that afflicts the world. The essential factor about both forms of pollution, of the sea and air, is that they are not confined to the borders of any one country. They are international, and ultimately it is only international agreement and action that can come to grips with the problem. The United States, after a belated start, now leads the world in the strictness of its environmental laws. But what of the American company drilling in the waters off Nigeria, or transporting oil from a Middle East field to a European market in a Liberian-flag tanker? The problem will eventually be solved when there is no more oil left, but by that time it will be far too late.

The development of various energy sources in the past has always been based on economics, but curiously it has also coincided with environmental needs. Wood was the traditional fuel, used widely by American industry even up to the end of the last century. Coal was discovered in time to prevent the complete destruction of the world's forests. But coal itself caused a growing pollution problem that was largely solved by the discovery of a much cleaner fuel, mineral oil. The smokeless London of today is cleaner than it ever was in Victorian times. In both cases the new fuels were developed on economic grounds, because they were both cheaper and more efficient than their predecessors. Now oil itself poses pollution hazards of a different nature. It might be cleaner than coal to burn as a boiler fuel, but coal will not pollute water if dumped at sea. Again there is another less harmful alternative, natural gas, and other forms of energy on the horizon such as the fast-breeder nuclear reactor and the battery-powered automobile that offer solutions for the future. But the world's fast-growing population and consequent surging energy demand do not allow time to rely solely on economic trends to bring these about, although even here they are helping as the cost of finding oil goes up and the producing countries demand ever higher payments for their exports. In the complex world of today it is only government intervention in the first place, and international action in the second, that can bring about the measures necessary to meet the energy demand and at the same time safeguard the environment. Private industry might argue that it has a highly responsible attitude to these needs and can best cope if left alone. This might be true about most of the major oil companies, but they no longer control the situation as they did in the past. In a competitive world the price for free enterprise is that some companies simply cannot afford conservation and must cut environmental corners to survive. It is equally unfair to expect some companies to foot the bill for conservation practices while others can ignore it if they wish and thus get the better of their competitors. This was clearly shown in the development of the North Slope where, although certain companies had a responsible attitude towards environmental needs, others did not and the damage was done all the same. Only by establishing legal ground-rules for all could this have been avoided.

On the North Slope in the winter of 1969–70, although the first rumblings of controversy could be heard, there was still optimism that the project would go ahead as planned, or at the most be delayed for a few months. Wells were being drilled at a faster rate than ever as the companies that had bought leases at the September sale eagerly sought to test their luck. The oil-crews themselves, phlegmatic men from Canada and Texas and Oklahoma, got on with their work with little concern about what outsiders in Washington and the big cities were saying. Charlie Wark, BP's Canadian lease superintendent who had been on the North Slope since January 1964, still kept a sprouting potato in a jar of water in his office as a reminder of things growing in the dead of an Arctic winter where nothing grows outside. He comes from a farming family.

'There's not much difference between here and Canada,' he said, 'except it's colder here. More than a hundred below sometimes. Not all the men who come here can stand it. Some of them leave after a few weeks. They make all kinds of excuses, but it's the isolation that gets them.'

Kayo Blackwell, a Texan tool pusher for the Parker Drilling Company at a lonely rig-site twenty miles from the main camp was one of those who stayed.

'They say you've got to live in Alaska three years to become a sourdough,' he said. The term comes from the kind of fermented dough that old timers used to make bread during their travels in the Yukon wilderness. A more ribald definition exists involving a polar bear, a bottle of whisky, the Yukon River and a squaw. 'But some of the guys who come here—it's such a shock for them that it only takes 'em two months to go sour on the country and run out of dough, so I guess you could call them sourdoughs too.'

George Mallard, another driller, was in Egypt immediately before coming to the North Slope.

'There's not much difference, except the weather. One camp is much like another anywhere in the world, until you step outside.'

Working twelve hours on and twelve hours off did not leave much time for anything except work, food and sleep, with an occasional movie thrown in. The most envied rigs on the North Slope were those with a good cook. The pattern of living in the camps fell into a well-ordered and even monotonous routine, broken only when a man finished his six-week stretch and went on three weeks leave, back to his family perhaps or the bright lights of Fairbanks.

'Maybe it's a dull life,' one driller said, 'but the money's good. And I can't stand the old woman for more than a few days at a time.'

'It's the cold that gets you,' said another. 'I took an orange over to the rig the other day. Just left it out for a couple of minutes, and it had frozen solid. When I threw it away it shattered into hundreds of pieces, like glass.'

It is taken for granted that something like a fresh orange should be readily available on the remote shores of the Arctic Ocean. These are the men who made the discovery of oil on the North Slope possible—the tool pushers, drillers, roughnecks and cat-skinners. Laconic, rather broody men who don't talk much but who have their own brand of humour and comradeship that, for all their grumbling and insistence on giving it all up, constantly brings them back to the oil camps, whether in Alaska, Texas, Indonesia or Libya. The country doesn't seem to matter very much. One camp is much like another. Some of them who are employed by contractors don't even know which oil company they are working for—and don't very much care either. The board decisions that brought them here, the calculations of the economists and the bargainings of the politicians are a world away. Even a new discovery causes little excitement. They know that most of the wells they drill will be dry—Suicide Number One is the usual name for a first well on an expensive lease bought as a gamble by some far-away group of financiers. It is all part of the endless job of drilling a few hundred feet, then pulling out the drill-pipe and unscrewing it section by section, changing the bit, then screwing each joint back again and lowering the pipe down the well to drill a few more hundred feet. By the time you've done that a couple of dozen times at a depth of 10,000 feet in a temperature of a hundred below, the end result doesn't mean a great deal. Whether a success or a failure, there's still another well to drill. There is just the job, and food, and sleep. Words like conservation and ecosystems belong to another world.

But that other world caught up with the oilmen of the North Slope. By the end of 1970, as the delay over the pipeline continued, companies began to abandon their plans for drilling until a decision was finally made. Whatever happened the pipeline would take about two years to build, so there was time enough then to drill for the oil it would carry. The oil boom went into cold storage. By mid-1972 there were only six wells operating where there might have been hundreds. Rigs were stacked or moved away to other areas. Most of the oilmen went with them, to drill in Texas or Indonesia or Libya. The same camps, just a different climate.

There is a legend told by the Eskimos of the North Slope. In the days when they lived by hunting, they always had the deepest respect for nature. They attributed a soul to all things, not only men and animals but hills and streams and stones as well. The weather was a man walking about the world. Winds came from holes in the sky which a shaman could sometimes close up and so make calm. The stars were human beings or animals escaped from the earth. If respect was not shown to an animal, even after it had been killed, it was believed that all the other animals would leave and the Eskimos would face starvation. For instance, a dipperful of fresh water was always poured into the mouth of a newly killed seal, and the skin of a polar bear was always hung up in the open for several days with certain valued implements to enable the bear's soul to depart peacefully in its own good time. Until that happened the animal was treated as an honoured guest who should not be offended.

'But there was once a mighty hunter who from pride decided it was not necessary for him to do these things. He had no respect for the souls of animals who offered themselves so readily to his spear and bow. And there came a time when all the animals went away from the land and there was a great hunger. The hunter travelled across the ice for many days in search of food and suffered many hardships. At last, weak and starving, he could go no further and lay down on the ground. And the Raven who was the creator of the earth came to him. The hunter pleaded for his life but there was no food available so the Raven turned him into a stalk of grass. He lay warm under the protection of the snow until summer. Then a family of mice came, eating and pulling up the grass, and he became frightened and wished he was a mouse. Immediately he changed into a mouse and was safe. But then an owl came swooping towards him and once more he was afraid and wished he could fly. And immediately he was changed into an owl and flew far away across the sea. But after a time he became tired and there was nowhere for him to land except on a piece of driftwood floating in the water. There he rested until he came near to a shore where he saw two men and they had food with them. He wished he was a man and immediately became one and landed on the shore. But because he was a stranger the two men, who were from another tribe, attacked and killed him. The moment he died the animals returned to the land of the Eskimos and they were hungry no longer.'

Bibliography

The State of Alaska, Ernest Gruening (Random House, New York)

Russian America, Hector Chevigny (The Cresset Press, London)

The Land Resources of Alaska, Hugh A. Johnson and Harold T. Jorgenson (University Publishers Inc., New York)

Alaska Natives and the Land, Federal Field Committee for Development Planning in Alaska (US Government Printing Office, Washington DC)

The Last of the Bush Pilots, Harmon Helmericks (Alfred A. Knopf, New York)

People of the Noatak, Claire Fejes (Alfred A. Knopf, New York)

Point Hope, James W. VanStone (University of Washington Press, Seattle)

Alaska—A Challenge in Conservation, Richard A. Cooley (University of Wisconsin Press, Milwaukee)

Minus 148 Degrees, Art Davidson (W. W. Norton & Co. Inc., New York)

Northwest Passage, William D. Smith (American Heritage Press, New York)

Index

Compiled by the author